INTERNATIONAL WILDLIFE ENCYCLOPEDIA

THIRD EDITION

Volume 1

AAR–BAR

Marshall Cavendish Corporation
99 White Plains Road
Tarrytown, New York 10591–9001

Website: www.marshallcavendish.com

Library of Congress Cataloging-in-Publication Data

Burton, Maurice, 1898-
    International wildlife encyclopedia / [Maurice Burton, Robert
Burton] .-- 3rd ed.
        p. cm.
    Includes bibliographical references (p.).
    Contents: v. 1. Aardvark - barnacle goose -- v. 2. Barn owl -
brow-antlered deer -- v. 3. Brown bear - cheetah -- v. 4. Chickaree -
crabs -- v. 5. Crab spider - ducks and geese -- v. 6. Dugong - flounder
-- v. 7. Flowerpecker - golden mole -- v. 8. Golden oriole - hartebeest
-- v. 9. Harvesting ant - jackal -- v. 10. Jackdaw - lemur -- v. 11.
Leopard - marten -- v. 12. Martial eagle - needlefish -- v. 13. Newt -
paradise fish -- v. 14. Paradoxical frog - poorwill -- v. 15. Porbeagle -
rice rat -- v. 16. Rifleman - sea slug -- v. 17. Sea snake - sole --
v. 18. Solenodon - swan -- v. 19. Sweetfish - tree snake -- v. 20. Tree
squirrel - water spider -- v. 21. Water vole - zorille -- v. 22. Index
volume.
    ISBN 0-7614-7266-5 (set) -- ISBN 0-7614-7267-3 (v. 1) -- ISBN
0-7614-7268-1 (v. 2) -- ISBN 0-7614-7269-X (v. 3) -- ISBN 0-7614-7270-3
(v. 4) -- ISBN 0-7614-7271-1 (v. 5) -- ISBN 0-7614-7272-X (v. 6) -- ISBN
0-7614-7273-8 (v. 7) -- ISBN 0-7614-7274-6 (v. 8) -- ISBN 0-7614-7275-4
(v. 9) -- ISBN 0-7614-7276-2 (v. 10) -- ISBN 0-7614-7277-0 (v. 11) --
ISBN 0-7614-7278-9 (v. 12) -- ISBN 0-7614-7279-7 (v. 13) -- ISBN
0-7614-7280-0 (v. 14) -- ISBN 0-7614-7281-9 (v. 15) -- ISBN
0-7614-7282-7 (v. 16) -- ISBN 0-7614-7283-5 (v. 17) -- ISBN
0-7614-7284-3 (v. 18) -- ISBN 0-7614-7285-1 (v. 19) -- ISBN
0-7614-7286-X (v. 20) -- ISBN 0-7614-7287-8 (v. 21) -- ISBN
0-7614-7288-6 (v. 22)
    1. Zoology -- Dictionaries.   I. Burton, Robert, 1941- . II.
Title.

QL9 .B796 2002
590'.3--dc21

                                                    2001017458

Printed in Malaysia
Bound in the United States of America

07 06 05 04 03 02 01   8 7 6 5 4 3 2 1

Brown Partworks
Project editor: Ben Hoare
Associate editors: Lesley Campbell-Wright, Rob Dimery,
Robert Houston, Jane Lanigan, Sally McFall, Chris Marshall,
Paul Thompson, Matthew D. S. Turner
Managing editor: Tim Cooke
Designer: Paul Griffin
Picture researchers: Brenda Clynch, Becky Cox
Illustrators: Ian Lycett, Catherine Ward
Indexer: Kay Ollerenshaw

Marshall Cavendish Corporation
Editorial director: Paul Bernabeo

Authors and Consultants

Dr. Roger Avery, BSc, PhD
(University of Bristol)

Rob Cave, BA (University of
Plymouth)

Fergus Collins, BA (University of
Liverpool)

Dr. Julia J. Day, BSc (University
of Bristol), PhD (University of
London)

Tom Day, BA, MA (University
of Cambridge), MSc (University
of Southampton)

Bridget Giles, BA (University of
London)

Leon Gray, BSc (University of
London)

Tim Harris, BSc (University of
Reading)

Richard Hoey, BSc, MPhil
(University of Manchester),
MSc (University of London)

Dr. Terry J. Holt, BSc, PhD
(University of Liverpool)

Dr. Robert D. Houston, BA, MA
(University of Oxford), PhD
(University of Bristol)

Steve Hurley, BSc (University of
London), MRes (University of
York)

Tom Jackson, BSc (University of
Bristol)

E. Vicky Jenkins, BSc (University
of Edinburgh), MSc (University
of Aberdeen)

Dr. Jamie McDonald, BSc
(University of York), PhD
(University of Birmingham)

Dr. Robbie A. McDonald, BSc
(University of St. Andrews), PhD
(University of Bristol)

Dr. James W. R. Martin, BSc
(University of Leeds), PhD
(University of Bristol)

Dr. Tabetha Newman, BSc, PhD
(University of Bristol)

Dr. J. Pimenta, BSc (University of
London), PhD (University of
Bristol)

Dr. Kieren Pitts, BSc, MSc
(University of Exeter), PhD
(University of Bristol)

Dr. Stephen J. Rossiter, BSc
(University of Sussex), PhD
(University of Bristol)

Dr. Sugoto Roy, PhD (University
of Bristol)

Dr. Adrian Seymour, BSc, PhD
(University of Bristol)

Dr. Salma H. A. Shalla, BSc, MSc,
PhD (Suez Canal University,
Egypt)

Dr. S. Stefanni, PhD (University
of Bristol)

Steve Swaby, BA (University of
Exeter)

Matthew D. S. Turner, BA
(University of Loughborough),
FZSL (Fellow of the Zoological
Society of London)

Alastair Ward, BSc (University
of Glasgow), MRes (University
of York)

Dr. Michael J. Weedon, BSc, MSc,
PhD (University of Bristol)

Alwyne Wheeler, former Head
of the Fish Section, Natural
History Museum, London

FOREWORD

WILDLIFE EXISTS WELL beyond the everyday reach of most people. Zoos and aquariums provide a glimpse of the wild, and documentary films bring dramatic images home from the field. Some people, like adventurers and scientists, with keen powers of observation and seemingly limitless patience report on species in natural environments that few people will ever visit. This third edition of the much-loved *International Wildlife Encyclopedia* organizes the ever-growing body of scientific information on animals for students and general readers eager for access to the details of a particular animal's way of life or for the enjoyment that can be found in reports on the wonders of nature.

In keeping with the venerable tradition established by the encyclopedia's first editors, Maurice Burton and Robert Burton, this thoroughly revised and updated edition includes nearly 1,200 articles providing a wealth of information on thousands of species from amoebas to zebras. A team of recognized experts on the physical features and ways of life of animal species has ensured that the encyclopedia is both accurate and up-to-date. Newly selected, color illustrations accompany every entry, and for the first time, color maps appear in most entries to illustrate discussions of species population, range and prospects for survival.

In addition to maintaining the high standard of lively, well-written accounts of the appearance and behavior of individual species, this edition introduces fifteen new articles on biomes and habitats, focusing on the interrelationship between survival and environment. More than thirty guidepost entries now single out some of the most popular animal groups for broad overviews of family relations. Including easy-to-read family trees, these guidepost articles also direct readers to hundreds of more detailed articles on related species. Special attention is given throughout these volumes to the taxonomic relations and systematic naming of species in order to help readers understand how scientists have classified genetic diversity among animal families.

Species from every branch of the animal world, every region of earth's landmasses and all oceans find representation in this set of extensively illustrated articles. Numerous animals not covered in entries of their own can be found referenced in various indexes based on common and scientific names, geographical location and behavior. The success and popularity of previous editions of the *International Wildlife Encyclopedia* and the considerable anticipation surrounding the release of this revision illustrate the enduring interest and enjoyment in expanding our knowledge and understanding of wild animals.

READER'S GUIDE

FACT FILE COLORS

▲ *Mammal*

▲ *Bird*

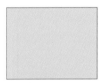

▲ *Reptile*

▲ *Amphibian*

▲ *Fish*

▲ *Invertebrate*

SPECIES ARTICLES

The *International Wildlife Encyclopedia* is arranged in alphabetical order by each animal's commonly used English name. The reader can easily locate an animal without referring to the index. Sometimes a species will have several common names, which may vary from country to country, or it may have locally evolved names. Alternative common names are given in the main text of each entry and are listed in the accompanying fact file. They also appear in the index.

Related animals are often grouped together in a multiple entry; for example, the "Alligator" entry includes both the American and Chinese alligators, while the "Angelfish" entry covers more than 150 species of angelfish. Sometimes, however, related animals are discussed in separate entries; for instance, the mandrill is a species of baboon and belongs to the baboon genus, but the "Mandrill" entry is found under "M" while the rest of the baboons can be found under "B."

FACT FILES

Fact files appear on the second page of each entry, or on the right-hand side of one-page entries. They provide an easy-to-read summary of the most important characteristics of each animal or group of related animals. Some fact files deal with a single species; others describe a number of closely related species. Each fact file is color-coded according to whether the animals it describes are mammals, birds, reptiles, amphibians, fish or invertebrates.

At the top of each fact file, a box provides the relevant scientific classification, including phyla, subclasses, suborders, superfamilies, subspecies and so forth, as needed. Below this, a second box lists the following information, where appropriate: alternative English names, weight, length, distinctive features, diet, breeding behavior, life span, habitat, distribution and status (see below for an explanation of status information). Imperial measurements are given, with the metric equivalents in parentheses.

MAPS

Most entries include a map showing the range of the species. There are some exceptions; for example, some invertebrates have such an enor-mous range spanning so much of the world that a map would not be of any help to the reader. Where many related species inhabit an area, the map may refer to more than one species.

GUIDEPOST ARTICLES

Throughout the encyclopedia there are guidepost articles that provide additional information about important groups of related animals such as bears, dolphins and sharks. The pages on which guidepost articles appear are all yellow for ease of reference.

Guidepost articles include a general description of the group's physical appearance, diet, habitat preferences, lifestyle and special adaptations as well as illustrations of species not covered elsewhere in the encyclopedia. The taxonomy, or scientific classification, of each group of animals is illustrated in two ways. A color-coded panel on the first page of the article lists the group's scientific (Latin) names, and where appropriate gives the total number of species contained in that group. A family tree then illustrates the precise relationships between the different orders, families, genera and species. At the end of each guidepost article, a "See also" reference directs the reader to relevant species entries elsewhere in the encyclopedia

BIOME AND HABITAT ARTICLES

Animal and plant species do not exist in isolation but together make up complex communities that are found in numerous microenvironments, habitats and biomes. A biome is a major ecological community type such as tropical rain forest or desert. The *International Wildlife Encyclopedia* includes articles on all of the world's important biomes and habitats. As with guidepost articles, all biomes and habitats articles have been printed on yellow pages to help the reader locate them.

For each biome or habitat the worldwide distribution, physical structure and climate are explained. Examples of typical animal and plant residents are described and illustrated. Other relevant issues such as biodiversity, food chains, food webs and conservation pressures are also discussed. Most biome and habitat articles include a fact file summarizing vital information.

INDEX

There are several ways in the Index Volume (Volume 22) for readers to find the entry they require. A species can be found by looking up either its common name, its scientific name or the geographical area in which it lives. For example, the Adélie penguin appears in the Comprehensive Index under "Adélie penguin" and "Penguin;" it appears in the Index of Places under "Antarctica" and "Birds" and in the Index of Scientific Names under "*Pygoscelis adeliae.*"

Many subspecies, or races, of animals that do not have their own entry in the encyclopedia are also indexed and cross-referenced. For example, the grizzly bear (a subspecies of the brown bear found in North America) appears in the Comprehensive Index under "Grizzly bear" and "Bear," in the Index of Places under "North America" and "Mammals" and in the Index of Scientific Names under "*Ursus arctos horribilis.*"

The Index of Animal Behaviors enables readers to search for many species by looking up certain aspects of their lifestyle, such as their diet and habits; for example, whether they live in groups or are nocturnal (active at night).

Volumes 1 to 21 each include their own index to the species described in that volume.

ADDITIONAL MATERIALS

The Index Volume contains a glossary and a bibliography as well as special sections providing internet resources and resources for younger readers. There is also a list of wildlife refuges throughout the world, with special emphasis on those in North America, and lists of museums, zoos and wildlife organizations that readers can visit or contact for additional information.

STATUS OF SPECIES

At the bottom of each fact file under the heading "Status" is an indication of how common the particular species is. The main source used to determine the status of each species is the Red List of Threatened Animals compiled by the I.U.C.N. (International Union for Conservation of Nature), which is often known as The World Conservation Union. The I.U.C.N. is a worldwide alliance of governments, governmental agencies and nongovernmental organizations based in Switzerland. Its aim is to help and encourage nations to conserve their animals, plants, habitats and natural resources.

As of 2001, the I.U.C.N. Red List includes more than 5,200 species and subspecies from around the world that are threatened with extinction. Of all the species that have been assessed by the I.U.C.N., it is estimated that 25 percent of mammals are threatened, 11 percent of birds, 20 percent of reptiles, 25 percent of amphibians and 34 percent of fish. The categories from the I.U.C.N. Red List used in the *International Wildlife Encyclopedia* are as follows:

• **Critically endangered:** a species that is facing an extremely high risk of extinction in the wild in the immediate future.

• **Endangered:** a species facing a very high risk of extinction in the wild in the near future.

• **Vulnerable:** a species that is facing a relatively high risk of extinction in the wild in the medium-term future.

• **Near-threatened or low risk:** a species that is threatened but which suffers a low degree of risk and so does not satisfy the criteria for critically endangered, endangered or vulnerable.

Where species are not under any serious threat, a number of general descriptive terms have been used as follows: rare; uncommon; locally common; common; abundant and superabundant. With certain species not enough scientific data is available to make an accurate asssessment of their status; in these cases, the species status is listed as "not known."

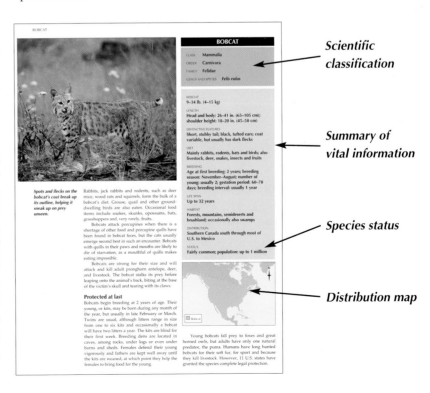

Scientific classification

Summary of vital information

Species status

Distribution map

TABLE OF CONTENTS

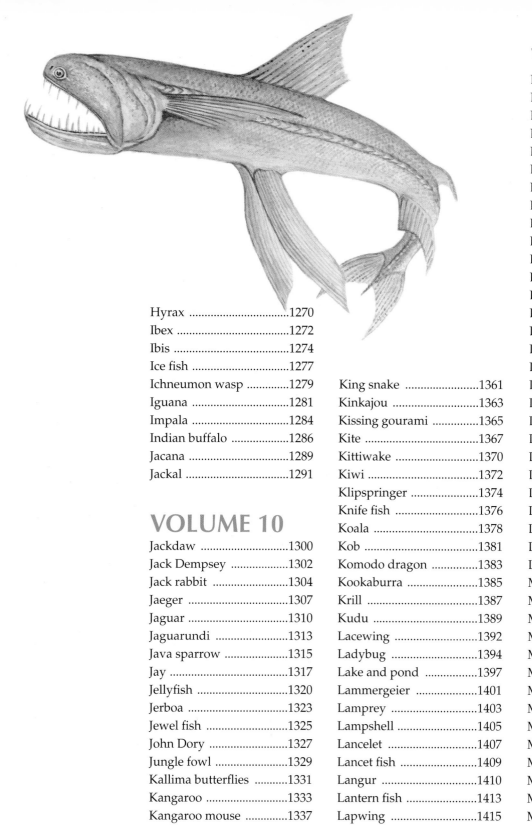

VOLUME 17

VOLUME 22

AARDVARK

T HE AARDVARK IS AN African mammal with a heavily built, bulky body. It may reach up to 6 feet (1.8 m) in length including its 2-foot (60-cm) tail, and stands 2 feet (60 cm) high at the shoulder. Its tough, gray skin is so sparsely covered with hair that it often appears naked except for areas on its legs and hindquarters. The aardvark's head is long and narrow, with donkeylike ears. Its snout bears a round, piglike muzzle and a small mouth, and its tail tapers from a broad root. Its feet have very strong claws, four on the forefeet and five on the hind feet.

Long burrows

The name aardvark is Afrikaans for "earth-pig." The animal has powerful limbs and sharp claws so that it can burrow quickly into the earth, as it might have to do if it is disturbed when away from its accustomed burrow. When digging, an aardvark rests on its hind legs and tail and pushes the soil back under its body with its powerful forefeet, dispersing the soil with its hind legs.

The burrow, usually occupied by a lone aardvark, averages 9–12 feet (2.6–3.6 m) long, but can be much more than this, perhaps up to 40 feet (12 m). Each burrow has a sleeping chamber at the end, big enough to allow the animal to turn around. The aardvark has several burrows and in some cases these can be miles apart. Abandoned burrows may be taken over by warthogs and other hole-dwelling creatures.

Years can be spent in Africa without seeing an aardvark, although it is found throughout Africa south of the Sahara. It does not live in dense forest, however, and is not found in the West African rain forests. Little is known of its habits as it is nocturnal and secretive, though it has to travel long distances in search of food.

Termite feeder

The aardvark's diet mainly comprises ants and termites, predominantly termites in the wet season and ants in the dry season. It has a muscular, gizzardlike stomach filled with grit or sand for crushing these hard-bodied insects. Aardvarks are important as seed dispersers as they disturb and disperse ant seed stores.

Aardvarks can rip through the walls of termite nests that would be difficult for humans to break down, even with a pick. Initially they tear a small hole in a wall using their powerful claws. The termites swarm at this disturbance, and the aardvarks then insert their slender, 1-foot (30-cm) tongue into the hole and pick the insects out. They are protected from termite attacks by very tough skin and the ability to close their nostrils, which are further guarded by a palisade of stiff bristles. Nonetheless, aardvarks are not immune to termites and stop feeding when their attacks become too intensive, moving to other termite mounds. As well as tearing open nests, aardvarks will seek out termites in rotten wood or while the insects are on the march. They also eat other soft-bodied insects and some fruit.

Breeding cycle

The female aardvark gives birth to a single young, twins being very rare. The young aardvark weighs around 4 pounds (1.8 kg) at birth and is born before or during the rainy season in its mother's burrow. It emerges after 2 weeks to accompany her on feeding trips. The young

Aardvark emerging from its burrow, South Africa. Aardvarks can burrow at high speed for considerable distances, up to 40 feet (12 m) in some cases.

AARDVARK

CLASS	**Mammalia**
ORDER	**Tubulidentata**
FAMILY	**Orycteropodidae**
GENUS AND SPECIES	***Orycteropus afer***

ALTERNATIVE NAME
Ant bear

WEIGHT
90–140 lb. (40–65 kg)

LENGTH
**Head and body: up to 4 ft. (1.2 m);
shoulder height: 2 ft. (60 cm);
tail: 2 ft. (60 cm)**

DISTINCTIVE FEATURES
**Large, donkeylike ears; piglike snout with
rounded muzzle; long, slender tongue;
large, muscular hindquarters; thick, tapering
tail; very tough, gray skin with sparse
covering of hair**

DIET
**Wet season: termites; dry season: ants; also
seeds, insect larvae and some fruits**

BREEDING
**Age at first breeding: 2 years; breeding
season: before or during rainy season;
gestation period: around 210 days; number
of young: 1; breeding interval: not known**

LIFE SPAN
Up to 10 years

HABITAT
**Grassland and open forests, always in sandy
soils that permit digging**

DISTRIBUTION
**Most of sub-Saharan Africa; absent from
West African rain forests**

STATUS
**Uncommon; scarce or extinct in many
agricultural areas**

Aardvark

moves with her from burrow to burrow, and is dependent on her for 6 months or more, until it is able to dig a burrow for itself.

Digs to escape predators

Although predation is rare, the aardvark may be preyed upon by humans, hunting dogs, lions, cheetahs and leopards. When suspicious it sits up kangaroolike on its hindquarters, supported by its tail, the better to detect danger. If the danger is imminent it runs to its burrow or digs a new one. When cornered, it fights back by striking with its tail or feet, even rolling on its back to strike with all four feet together. However, flight and, above all, superb digging ability are the aardvark's first lines of defense.

A creature on its own

One of the most remarkable things about the aardvark is the difficulty zoologists have had in finding it a place in the scientific classification of animals. At first it was placed in the order Edentata (meaning "toothless") along with the armadillos and sloths, simply because of its lack of front teeth (incisors and canines). Now it is placed by itself in the order Tubulidentata (meaning "tube-toothed"), so called because of the fine tubes radiating through each tooth. These teeth are in themselves remarkable, for they have no enamel, being covered instead by cementum.

Recent research has found that aardvarks can instead be grouped into a new taxon, Paenungulates, which includes the elephant shrews, hyraxes, elephants and dugongs.

*Aardvarks rip open
ant and termite nests,
lapping the insects
out with their long
slender tongues.*

AARDWOLF

A young female aardwolf. Due to their striking resemblance to hyenas, aardwolves are mistakenly persecuted by humans.

THE AARDWOLF IS A member of the hyena family, but differs from the true hyenas in that it has five instead of four toes on its front feet, relatively larger ears and a narrower muzzle. It is of a smaller, lighter build than the spotted hyena, *Crocuta crocuta*, its body being up to 2½ feet (75 cm) long with a 1-foot (30-cm) tail. It weighs 15–25 pounds (6.8–11.3 kg). The aardwolf's coat is yellow gray with black stripes, except for its legs, which are black below the knee. Its muzzle is black and hairless, and it has powerful jaws and strong canines that it uses for defending its territory. Its tail is bushy and black-tipped, and the hair along its back and neck is long. This ridge of hair usually lies flat, but when the animal is frightened it erects the hair on the neck, or, in extreme cases, along its whole back. The name aardwolf is Afrikaans for "earth-wolf".

Separated populations

There are two separate populations of aardwolves, one found throughout southern Africa, the other ranging through eastern Africa as far north as Sudan. These populations are separated by a 930-mile (1,500-km) gap. The aardwolf is not common in any part of its range, although the southern African population is now thought to be stable. It is found most frequently on grassland, sandy plains and open, scrubby forest, but is rarely seen since it is a nocturnal animal. The aardwolf spends the day lying up in rock crevices or in burrows. The burrow consists of two or more sharply winding tunnels, 25–30 feet (7.6–9 m) long, leading to a sleeping chamber about 3 feet (90 cm) in diameter.

A hyena that eats termites

The aardwolf is a very specialized carnivore and lives almost exclusively on snouted harvester termites, although it does take some other insects. It lacks claws strong enough to tear open termite nests and, as a result, is limited either to licking up the insects from the ground's surface, or digging them out of soft soil. Its long, tacky tongue is well adapted for sweeping the insects up, however, being shaped like a spatula with pronounced papillae (small protuberances) on the upper surface. Its stomach is also highly muscular and gizzardlike, another adaptation to its diet of insects. The aardwolf's speed and

AARDWOLF

CLASS	**Mammalia**
ORDER	**Carnivora**
FAMILY	**Hyaenidae**
GENUS AND SPECIES	***Proteles cristatus***

ALTERNATIVE NAME
Werewolf

WEIGHT
15–25 lb. (6.8–11.3 kg)

LENGTH
**Head and body: 2–2½ ft. (60–75 cm);
tail: 1 ft. (30 cm)**

DISTINCTIVE FEATURES
**Resembles spotted hyena, but has smaller,
lighter build; large ears; narrow muzzle;
yellow-gray coat with striped back and
bushy tail; black lower legs and tip to tail;
ridge of long hair along back and neck**

DIET
**Almost exclusively snouted harvester
termites; rarely other insects, mice and
bird nestlings**

BREEDING
**Age at first breeding: not known; breeding
season: winter; number of young: 1 to 4;
gestation period: 60–90 days; breeding
interval: 1 year**

LIFE SPAN
Up to 15 years in captivity

HABITAT
Grassland and open, scrubby forest

DISTRIBUTION
**Two populations: South Africa north to
southern Angola and Zambia, and eastern
Africa from central Tanzania to Sudan**

STATUS
**Rare, but southern African population is
thought to be stable**

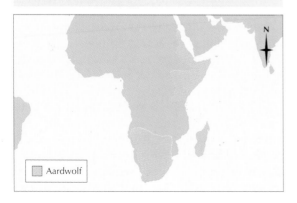

Aardwolf

efficiency in feeding is demonstrated by the fact that each animal is able to take up to 250,000 termites each night, perhaps 100 million a year.

When insects are in short supply, the aardwolf may turn to other prey such as mice, small birds and the eggs of ground-nesting birds. It has been reported to eat carrion, but in such cases it is more likely to be feeding on the beetles and maggots within the carcass.

Breeding cycle

Aardwolves form mating pairs and each pair establishes a large territory of ½–1 square mile (1.3–2.6 sq km). The territory consists of several termite patches enclosed by boundaries, and it is defended aggressively by both the male and the female. Males are reported to desert females if the females are promiscuous. Mating takes place in winter, usually in July. A litter of one to four young is born after a gestation period of 60–90 days, in October or November. The young are born blind and are cared for by both parents. One litter is produced each year.

Mistaken persecution

The aardwolf's main threat is human beings, who often mistake it for a hyena. Hyenas are themselves hunted because a bounty is paid for killing them in many areas. Despite being protected in some places, aardwolves occasionally suffer persecution in farming country, both through mistaken identity and the idea that they take poultry. Their main natural predators are probably jackals, which attack their young.

When acting in defense of themselves or their territory, aardwolves put up a good fight with their long canine teeth and are reported to eject a strong, musky fluid from their anal glands.

Aardwolves differ from hyenas in their highly specialized diet, which consists almost entirely of snouted harvester termites.

ABALONE

ABALONES COMPRISE ABOUT 100 species of mollusks related to the limpets. Also known as the ormer, sea ear or earshell, the abalone somewhat resembles a terrestrial snail. Its body is little more than a muscular foot with its head at one end, bearing a pair of eyes and sensory tentacles. Its body is also fringed with tentacles. Along the side of its shell is a line of between 4 and 10 holes through which water is exhaled after it has been drawn in under the shell and over the gills to extract oxygen. New holes are formed as the shell grows forward, while the old holes become covered over.

Some species are among the largest shellfish in the world. They range in size from just ⅖ inch (1 cm) long to the red abalone, *Haliotis rufescens*, of California, which is up to 10 inches (25 cm) across and may weigh up to 2⅕ pounds (1 kg).

Distribution and habits

Abalones are to be found in many parts of the world, including the coastal zones of the Mediterranean region, Africa, Australia, New Zealand, the Pacific islands and western North and South America. One of the smallest and rarest species, *Haliotis pourtalese*, is found off Florida in the United States. It is known mainly from specimens washed up on the shore, as it lives at depths of 330–3,600 feet (100–1,100 m). It is thus the deepest-living of all abalones. Other species live between the extreme low-water mark and a depth of about 65 feet (20 m). They are found along rocky shores where there is no sand to clog the gills, or in rocky pools large enough not to be heated too quickly by the sun. The only other exception is the black abalone, *Haliotis cracherodii*, which lives in the splash zone where waves breaking against rocks alternately cover and expose it.

Unlike their limpet relatives, abalones have no spot on a rock to which they always return after feeding. They simply hide up in a crevice or under a rock, coming out at night. When disturbed, an abalone grips the rock face, using its foot as a suction pad. The two main muscles of the body exert a tremendous force but, unlike a limpet, the abalone cannot bring its shell down over the whole of its body and the edge of its foot, with its frill of tentacles, is left sticking out.

Abalones move in the same way as limpets and terrestrial snails. Waves of muscular contraction pass along the foot, pushing it forward. However, they have an almost bipedal (two-footed) gait, with alternate waves of movement passing down either side of the foot. The rate of travel is rapid for a shellfish and speeds of over 16 feet (5 m) per minute have been recorded.

Many-toothed tongue for feeding

Abalones crawl over rock faces, browsing on seaweed that they seek out with their sensitive tentacles. Their favorite foods are the delicate red weeds and green sea lettuces, although some of the larger species also eat kelp. Young abalones eat the forms of life that encrust rocks, such as the coral-like plant *Corallina*. They scrape up food and chew it into small pieces using the rasp-like action of the radula, a tongue made up of large numbers of small, chalky teeth.

Millions of eggs

Unlike some mollusks, the sexes are separate in the abalone. The germ cells, or gametes, are shed directly into the sea, causing a great deal of wastage. Hence, a female will release between 1 and 10 million eggs, and the sea around a male turns milky over a radius of 3 feet (90 cm) when he sheds his milt (sperm-containing fluid). However, wastage is reduced because the female is stimulated to release her eggs only in the presence of male sperm.

The fertilized eggs float freely in the sea until they hatch a few hours later as minute trochophore larvae. These swim around by means of a band of hairlike cilia. Within a day, each trochophore develops into a veliger, a miniature version of the adult, complete with shell but

The underside of an abalone, Haliotis tuberculata, *showing its broad foot and the fringe of tentacles it uses to find food.*

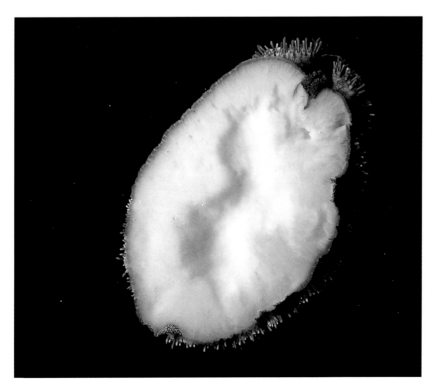

ABALONES

PHYLUM **Mollusca**

CLASS **Gastropoda**

SUBCLASS **Prosobranchea**

ORDER **Archaeogastropoda**

FAMILY **Haliotidae**

GENUS AND SPECIES **About 100 species, including red abalone, *Haliotis rufescens*; green abalone, *H. fulgens*; and black abalone, *H. cracherodii***

ALTERNATIVE NAMES
Ormer; sea ear; earshell

LENGTH
⅖–10 in. (1–25 cm), depending on species

DISTINCTIVE FEATURES
Row of 4 to 10 holes along part of shell; inside of shell has mother-of-pearl layer

DIET
Seaweed, especially green and red seaweed; also kelp (some larger species only)

BREEDING
Age at first breeding: may be 6 or 7 years; breeding season: varies with species and geographical location; number of eggs: 1 million to 10 million; hatching period: a few hours; breeding interval: not known

LIFE SPAN
Most species: 7 or 8 years; much longer than this in some species

HABITAT
Rocky shores and coral reefs; shallow waters, rarely deeper than 65 ft. (20 m)

DISTRIBUTION
Mainly tropical, subtropical and warm temperate areas; largely absent from western Atlantic

STATUS
Varies with species

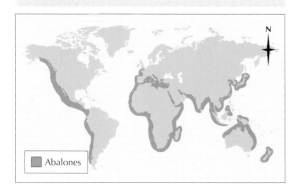

Abalones

still with the band of cilia. Two days later it loses the cilia, sinks to the bottom and starts to develop into an adult, a process that usually takes several weeks.

Many predators

The free-swimming larvae have advantages in that they are the means by which the otherwise rather sedentary abalones can spread, but they are very vulnerable to predators. They are eaten in their millions by plankton-eating fish such as anchovies and herrings.

Adult abalones are also preyed upon by fish, seabirds, sea otters, crabs and starfish. Their only protection lies in their tenacity in clinging to rocks and the protective camouflage of the shell and foot. This camouflage is improved by the seaweed and sedentary animals that settle on the shell. In addition, when young abalones feed on red weeds, their shells become red. Likewise, if they eat green seaweeds, they become green.

Aquaculture

The abalone is a popular food for many people. This, along with the ease with which it can be collected, has led to stocks being severely depleted in many parts of the world. In California, the center of the United States abalone industry, only strict laws have prevented its extinction. In Japan, meanwhile, many millions of juveniles are released into the wild each year, having been reared for 1 year or so, in order to maintain stocks. There are also many small-scale farming ventures worldwide, rearing abalones to full size for the edible market. Their price is fairly high, but it is still difficult to make such aquaculture economically viable because abalones require at least 4 or 5 years to reach a marketable size.

The abalone is protected from its many predators by the camouflage of its shell, and by the tremendous force with which it grips onto rocks.

ACCENTOR

T HERE ARE 13 SPECIES of these small, rather sparrowlike birds, forming a single genus and family. They differ from sparrows in having slender and finely pointed bills and a well-developed tenth primary wing feather. Accentors are generally regarded as being related to thrushes or warblers. The most common European species, the dunnock or hedge accentor, *Prunella modularis*, is rather featureless, yet identifiable by its gray breast, neck and head, and dark brown wings. Its song, a hurried jingle similar to that of the European wren, *Troglodytes troglodytes*, can be heard virtually all year round. The Siberian accentor, *P. montanella*, is so named because it breeds in northern Siberia. A few inviduals reach Alaska in the autumn, having strayed off course en route to their winter quarters in eastern China and Korea.

Unobtrusive birds

Accentors are found in Europe and Asia. The dunnock can be seen all over Europe, except in parts of the far north and in Iceland, Hungary, Romania and Ukraine. It is common in Britain wherever the habitat is suitable, except in the north, where is it seem more rarely. The alpine accentor, *P. collaris*, can be found on mountain ranges from Spain east to Japan, extending as far south as North Africa.

The typical habitat of most accentors is in mountainous regions, often well above the tree line and up to the snow line. The Himalayan accentor, *P. himalayana*, is found breeding as high as 17,000 feet (5,182 m) above sea level, and one subspecies of the alpine accentor breeds up to 18,500 feet (5,640 m). However, most species breed in the scrub vegetation at rather lower levels. Some species are hardy enough to spend the winter at high altitudes, but others migrate downward. The remainder live in forests. The dunnock is found in many kinds of habitats, but especially in gardens, hedgerows and scrubland.

Accentors are quiet and unobtrusive, remaining close to the ground in the undergrowth. If flushed they fly low and in undulating fashion, seeking cover. On the ground they proceed by leisurely hops or a kind of creeping walk, with the body almost horizontal. The wings are often flicked. This is most noticeable in the dunnock during courtship and has earned it the traditional name of shuffle-wing.

Several species in the accentor family, Prunellidae, tend to live together in small flocks. The dunnock, however, is usually solitary outside the breeding season, coming together in small groups only for feeding and a peculiar wing-flicking display. There is little migration: movements are mainly just from higher to lower ground and from far north to south as temperatures drop in the fall. Vagrant alpine accentors, however, have reached the Faroe Islands, located between Iceland and Denmark, and Siberian accentors have turned up in Alaska.

Insects in summer, seeds in winter

During the summer months accentors eat spiders and insects. In winter they live almost entirely on seeds and berries, even picking them out of

The dunnock, with its characteristic gray head and brown wings, is typically inconspicuous.

DUNNOCK

CLASS	**Aves**
ORDER	**Passeriformes**
FAMILY	**Prunellidae**
GENUS AND SPECIES	***Prunella modularis***

ALTERNATIVE NAMES
Hedge accentor; hedge sparrow; shuffle-wing (archaic)

WEIGHT
½–¾ oz. (16–25 g)

LENGTH
Head to tail: about 6 in. (15 cm); wingspan: 7–8 in. (19–21 cm)

DISTINCTIVE FEATURES
Slender bill; gray head, neck and breast; brown upperparts, streaked with black; off-white belly

DIET
Summer: spiders and insects such as beetles, butterflies and flies; winter: seeds and berries

BREEDING
Age at first breeding: 1 year; breeding season: March–June; number of eggs: usually 4 to 6; incubation period: 12–13 days; fledging period: 11–12 days; breeding interval: usually 2 broods per year

LIFE SPAN
Up to 9 years

HABITAT
Woodland clearings, spruce plantations and scrub; also gardens, parks and hedgerows in Britain

DISTRIBUTION
Europe and western Asia, except far north, Iceland and parts of eastern Europe; also in parts of North Africa

STATUS
Common

Dunnock

animal droppings. Accentors have a finchlike crop and muscular gizzard. They swallow grit to help break up the seeds.

This alpine accentor, a larger bird than the dunnock and with white markings on its wings, is perched at a height of 14,000 feet (4,270 m) in the mountains of Nepal.

Complex mating system

The dunnock has a complex mating system. We can see simple pairs (monogamy), a male with two females (polygyny), a female with two males (polyandry) and two males sharing two or more females (polygynandry). Both sexes are trying to get the best deal by maximizing reproductive output and minimizing parental care. The resulting mating system usually depends on the strength of the competing birds.

The males sing from rocks or low bushes, sometimes making short, larklike song flights. The male dunnock does not participate in nest-building or incubation. The female makes a cup-shaped nest in a rock crevice or in a shrub, out of leaves, twigs, moss and grasses, sometimes with a few feathers. Very occasionally, dunnocks use a lot of feathers for the lining of the nest. Sometimes the old nest of a Eurasian blackbird or barn swallow is used. About five dark blue eggs are laid, which the hen incubates for about 12 days, leaving the nest only to feed. In accentor species other than the dunnock, the male shares in nest-building and incubation. The young are fed by both parents and fledge in about 12 days. Those of the alpine accentor sometimes leave the nest before they can fly.

Singing in all seasons

The male accentor's song is neither very loud nor distinctive, but it is persistent. The dunnock has a short, high-pitched warbling song that is heard during all seasons. It is most constantly and vigorously repeated when the bird is excited, as when two rival males meet or when male and female are courting. The dunnock's main song period is during its breeding season, yet courtship begins in December and its song gains vehemence at a time when most birds are relatively quiet.

ADDAX

Habitat destruction and hunting have reduced the addax to a few small populations in West Africa. Only one viable herd is thought to remain in the wild.

THE ADDAX IS A SINGLE species in the antelope family, closely related to the oryxes. It is also known as the screwhorn antelope and differs from most antelopes in the absence of facial glands and in its large, square teeth, which are more like those of cattle.

An adult male stands some 3½ feet (1.1 m) at the shoulder and weighs up to 420 pounds (190 kg). The addax's coat varies in color with the season. In winter it is grayish brown with white hindquarters, underparts and legs, while in summer its body becomes sandy or almost white. Its head is white and distinctly marked with brown and black patches to form a white "X" over the nose. The addax has a tuft of long, black hairs between its horns, and has a short mane. Its tail is short and slender, tipped with a tuft of hair.

Both sexes bear horns, the female's being somewhat thinner than the male's. The horns are similar to those of the oryxes, but curve out from the base and spiral back over the neck. The horns might reach a length of nearly 3 feet (90 cm), measured in a straight line from base to tip.

Increasingly rare

At one time the range of the addax extended across the Sahara Desert from the Atlantic coast to Egypt, particularly in the sand-dune areas. The ancient Egyptians are known to have kept it in at least semidomestication because pictures in a tomb dating from 2500 B.C.E. show addax and other antelopes wearing collars and tethered to stakes. It also seems that the number of addax a person owned was a status symbol at the time, an indicator of their wealth and position.

More recently, addax were to be found from Algeria to the Sudan, but never farther south than a line drawn roughly from Dakar, Senegal, to Khartoum in the Sudan. Now these animals are much more restricted and are becoming increasingly rare. They are still found in Niger and Chad, with small populations in Mauritania and Mali. One recent study estimated there to be only one viable herd, of some 50 to 200 individuals, remaining in the wild in northeastern Niger. Addax are now extinct in Algeria and it is thought that they may also be extinct in the Sudan and the western Sahara.

Details of captive addax are recorded in a book held at San Diego Zoo in North America. During 1966 and 1967, in seven zoos throughout the world, 33 offspring were bred and in 1974 there were 146 captive addax around the world. By 1988 this figure had increased to 409 animals.

Threats to survival

There are two main reasons for the decline in wild populations. First, the addax's habitat is being destroyed by the opening up of desert areas for commercial projects. At the same time the sparse desert vegetation is being destroyed by herds of domestic goats.

Secondly, addax continue to be killed by hunters for their horns and hides. In 1966 they were given formal protection, but this has not been easy to put into practice. The addax is slow by comparison with other antelopes, so it falls easy prey to humans hunting with dogs or in motorized vehicles. Nor is it difficult to pursue an addax to exhaustion, for it will panic and use up its energy in a blind attempt to maintain a high speed. Modern hunters in cars can exhaust an addax in less than 10 minutes.

Adapted for desert living

One factor that might enhance the addax's chances of survival is its adaptation to a desert habitat. Its hooves are short and widely splayed, enabling it to travel over the sand in the long journeys that are necessary for it to cover large areas in search of scanty supplies of food. Moreover, the addax is able to survive in the very depths of the desert, where conditions are so extreme that no other warm-blooded animal can survive permanently. Although it can drink large quantities of water at a time, the addax is nevertheless able to survive without any free water

ADDAX

CLASS	**Mammalia**
ORDER	**Artiodactyla**
FAMILY	**Bovidae**
SUBFAMILY	**Hippotraginae**
GENUS AND SPECIES	***Addax nasomaculatus***

ALTERNATIVE NAMES
White antelope; screwhorn antelope

WEIGHT
220–420 lb. (100–190 kg)

LENGTH
Head and body: 6 ft. (1.8 m);
shoulder height: 3–3½ ft. (0.9–1.1 m);
tail: 6–9 in. (15–23 cm)

DISTINCTIVE FEATURES
Large antelope; spiralled horns (both sexes);
dark mane; in winter coat is grayish brown
with white hindquarters, underparts and
legs; in summer whole body becomes
sandy or white; markings on face form
white "X" over nose

DIET
Desert vegetation

BREEDING
Age at first breeding: 2–3 years (female);
breeding season: births in winter or early
spring; number of young: usually 1;
gestation period: 255 days

LIFE SPAN
Up to 25 years in captivity

HABITAT
Sandy and stony desert

DISTRIBUTION
Chad, Mali, Mauritania and Niger; may be
extinct in Sudan and western Sahara

STATUS
Critically endangered and in danger of
total extinction in wild

Addax

almost indefinitely. It is able to obtain sufficient water either from succulent vegetation or from the dew that condenses on plants.

Little known antelope

The addax's habits are not well known, owing to the thinly spread and inaccessible nature of the surviving populations. It is also very wary and at the slightest alarm will dash off at a gallop. Typically, addax move about in small herds of between 4 and 20 animals, rarely more than 30, led by several adult males.

Almost nothing is known about the addax's breeding except that one calf is born, usually in winter or early spring. In captivity, at least, the first calf is born when the mother is 2 or 3 years old. The gestation period lasts 255 days.

Staple diet of grass

The movements of addax are related to the distribution of their food, which in turn is related to the weather. They are most likely to be found along the northern fringe of the tropical summer rains, moving north in winter as the Mediterranean trough system brings rain southward. They can tell where the rains have fallen by scenting from a distance where the vegetation has turned green.

Addax feed mainly on the *Aristida* grasses, perennials that are green throughout the year. Another favored food in southern areas is *Parnicum* grass. When feeding on the *Aristida* grasses addax crop all the blades to a level height. However, the outer, dried blades of *Parnicum* grass are not touched. The addax take only the fresh green blades, pushing their heads into the middle of the clump and gripping the growing stems. *Parnicum* seeds are also favored. As the seeds are present throughout most of the year and are rich in protein, they form a valuable item of the addax's diet.

A young addax, 3 or 4 weeks old. The female gives birth to a single calf in winter or early spring, but little else is known of the addax's breeding habits.

ADDER

THE ADDER IS A MEMBER of the viper family. It has a relatively stout body for a snake, and a short tail. The average male is 1¾ feet (54 cm) long, while the female is about 2 feet (60 cm). The record length is 2⅖ feet (81 cm). The adder's head is flat, broadening behind the eyes to form an arrowhead shape. Its color and body markings vary considerably, and the most distinctive of these is probably the dark, zigzag line down the back with a series of spots on either side. The adder's head also carries a pair of dark bands, often forming an "X" or a "V."

Distinguishing males and females

It is often possible to distinguish the sex of an adder by its color. Males are usually cream, yellowish, silvery, pale gray or light olive, with black markings. Females are more often red, reddish brown or gold, with darker red or brown markings. The male's throat is black, or whitish with the scales edged or spotted with black. Females, meanwhile, have a yellowish white chin sometimes tinged with red.

The adder ranges throughout Europe and across Asia to Sakhalin Island, north of Japan. Across its range the adder is usually seen in dry places such as sandy heaths, moors, rough grassland, clearings in deciduous and pine forest and sunny slopes. It often basks in the sun on hedge banks, logs and piles of stones.

Adders vary in color and body markings, but all have a dark, zigzag line down the back. They have the widest geographical distribution of any terrestrial snake.

Hibernates in groups

The adder's tolerance of cold allows it to live as far north as Finland, within the Arctic Circle. It escapes cold weather by hibernation, which starts when the temperature falls below 49° F (9° C). It emerges again when the air temperature rises above 46° F (8° C), but a cold spell will send it back into hibernation. The duration of hibernation depends, therefore, on climate: in northern Europe it may last up to 275 days, whereas in the south it may be as little as 105 days.

Unlike many other snakes, adders do not burrow but seek out crevices and holes where they lie up for the winter. The depth at which they hibernate depends, like duration, on the climate. They are found at greater depths where the winters are more severe. Very often many adders will be found in one den, or hibernaculum. As many as 40 have been found coiled up together. This massing together is a method of preventing heat loss, but it is not known how the adders come to congregate in the hibernacula, which are used year after year.

It is uncertain whether adders are nocturnal (night-active) or diurnal (day-active). Their eyes are typical of nocturnal animals in that they are rich in the very sensitive rod cells. Such eyes are effective at night, but during the day they need protection, and the adder's slit pupils cut down the intensity of light. Despite these adaptations, adders are often active during the day. Courtship and some hunting are definitely diurnal.

Rodent killer

The adder's main prey items are lizards, mice, voles and shrews. Young adders subsist at first on insects and worms. Larger victims are killed by a venomous bite, the effects of which vary with the size of the prey. A lizard, for example, will be dead within a few minutes. The adder's method of hunting is to follow its prey by scent, then attack it with a quick strike of the head. While the poison acts, the victim may have time to escape to cover, in which case the snake will wait for a while before following to eat its lifeless prey.

Dance of the adders

The mating period is from the end of March to early May. In the north of Europe the summer is too short for the eggs to mature in one year, so breeding takes place in alternate years.

At the beginning of the breeding season, there is a good deal of territorial rivalry between males, culminating in the "dance of the adders." Two males face each other with their heads held

ADDER

CLASS	**Reptilia**
ORDER	**Squamata**
SUBORDER	**Serpentes**
FAMILY	**Viperidae**
GENUS AND SPECIES	***Vipera berus***

ALTERNATIVE NAMES
Viper; northern viper

LENGTH
Head and body: usually up to 2 ft. (60 cm); tail: up to 2¾ in. (7 cm)

DISTINCTIVE FEATURES
Color and markings vary considerably and are different in male and female; both sexes have dark zigzag stripe running along back and dark "V" or "X" shape immediately behind head

DIET
Adult: mice, voles and shrews; also lizards. Young: insects and worms.

BREEDING
Age at first breeding: 4 years; breeding season: April–May; number of young: usually 10 to 14; hatching period: about 120 days; breeding interval: 2 years (north of range), 1 year (rest of range)

LIFE SPAN
Probably 10–15 years

HABITAT
Heathland, tundra and clearings in deciduous and pine forests; also meadow edges, rough grassland, hedgerows and roadside verges; occasionally suburban gardens

DISTRIBUTION
Throughout Europe and east across Asia to Sakhalin Island, north of Japan

STATUS
Common

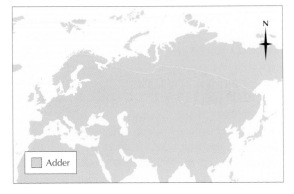

Adder

erect and the forepart of their bodies off the ground. They sway from side to side, then with bodies entwined each attempts to force the other to the ground. They do not bite each other. Finally one gives up and departs. The female, which is frequently waiting close at hand, will accept the victorious male if she is ready to mate.

Adders are ovoviviparous, that is, the eggs remain inside the mother's body until they are fully developed, and the young are born coiled up in a membrane that is ruptured by their convulsive movements. The young are born in August or September and the number ranges from 6 to 20, although 10 to 14 is most common. Each measures 6–8 inches (15–20 cm) in length. The young adders are immediately capable of independent existence.

Male adders fighting in prenuptial display. If the female is ready to mate, she will accept the victorious male.

Predators defy venom

Like the majority of animals, adders are most likely to flee if confronted with danger, and bite only if suddenly frightened. Humans are their chief enemy. The killing of adders on sight has not led to any serious decline in their numbers, although nowadays increased urbanization is destroying their habitat.

Undoubtedly, many carnivores will take adders. Red foxes and Eurasian badgers kill them, and they have been found in the stomachs of pike and eels. However, their greatest adversary is the hedgehog. One reason is that it can tolerate large doses of venom without harm. Its method of killing is to bite the adder, then curl up leaving nothing but a palisade of spines for the snake to strike at. It repeats the process of biting and curling until the snake is dead.

ADÉLIE PENGUIN

PENGUINS ARE FOUND IN Antarctica, but are not, as is often thought, restricted to the frozen land and sea. No penguins are found in North American waters, but various species do live around the coasts of South America, Africa and Australasia. These penguins do not usually go far north but stay where the sea is quite cool. However, the Adélie penguin, with the emperor penguin, is one of the species that is actually confined to the Antarctic continent and its neighboring islands.

The Adélie penguin, which stands about 2⅓ feet (71 cm) tall, is simply colored, with a white belly and black back and throat. The eye is distinguished by a surrounding circle of white. Like all penguins Adélies are flightless, gregarious birds. Superb swimmers, they are well adapted to life in water. Their wings have evolved into flippers, while their bodies are covered with a protective layer of blubber.

Antarctic environment
Adélie penguins spend the long winter on the edges of the frozen seas. At first sight it might seem surprising to find colonies of thousands of penguins in the apparently desolate wastes of the Antarctic. However, the Antarctic seas are teeming with life, especially with small shrimp-like creatures such as amphipods and krill on which the penguins, as well as seals and whales, feed. The Adélie penguin's diet also comprises some fish and squid.

Adélie penguins travel to their nesting grounds, or rookeries, in September or October. After feeding through the winter they are in prime condition, their bodies padded with blubber and their feathers sleek and glossy. At first the groups consist of a dozen or so birds, but they increase in numbers until streams of penguins are flowing in towards the rookeries, where they nest in their thousands.

Return to old nests
At the rookery, which is usually situated on a rocky headland, each penguin searches for its old nest, or if it is breeding for the first time, finds an empty space. The males usually arrive first and they stand on the nest site, fighting off other males and waiting for their mates. They have a special display, called the ecstatic display, whereby the male points its bill to the sky, waves its flippers and utters a loud, braying call. This at once intimidates other males and attracts females.

When all the penguins have arrived and the pairs have formed, each nest site is exactly the same distance from its neighbors. This even spreading ensures that the eggs and chicks will not be disturbed. Occasionally a penguin does get in the way of its neighbors and a fight breaks out, the birds pecking and beating one another with their flippers.

Take turns to feed
When the snow melts, nest-building can begin. The male collects pebbles and takes them to the female. He drops each pebble at her feet and she uses them to build up a ring around herself on the nesting site. Usually the pebbles are laboriously collected from the beach, but the penguins miss no chance to steal from any unguarded nest.

Two white eggs, each 2 inches (5 cm) long, are laid in the nest of pebbles. The male broods them while the female goes back to the sea to feed. By this time she will not have eaten for 2–3 weeks. A fortnight later she returns and the male goes off to break his fast of some 6 weeks. During this time he will have lost almost half his body weight. The eggs hatch after 30 to 43 days.

While one parent is guarding the chicks the other collects food for them, returning with it stored in the crop where it is partly digested. Reaching the nest, the adult penguin opens its bill for the chick to take the food. The chicks grow rapidly, coming out from under the parents

Unlike many other penguins, Adélies spend much of their lives on the edges of the frozen seas around the Antarctic continent.

ADÉLIE PENGUIN

CLASS **Aves**

ORDER **Sphenisciformes**

FAMILY **Spheniscidae**

GENUS AND SPECIES ***Pygoscelis adeliae***

LENGTH
Head to tail: about 2⅓ ft. (71 cm)

DISTINCTIVE FEATURES
Streamlined body with short feathers; white underparts and black upperparts; white ring around eye; webbed feet; stiff flippers (modified wings)

DIET
Mainly krill; some fish and squid

BREEDING
Age at first breeding: normally 8 years, very occasionally 3 or 4 years; breeding season: September–February; number of eggs: 2; incubation period: 30–43 days; fledging period: 50–56 days; breeding interval: 1 year

LIFE SPAN
Not known

HABITAT
Cold seas and ice-free, rocky coasts

DISTRIBUTION
Antarctic continent and its neighboring islands

STATUS
Common; population: 2 to 2.6 million pairs

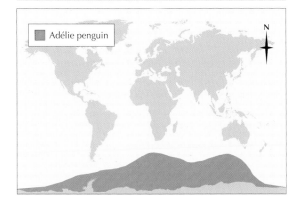

Adélie penguin

Adélie penguin breeding colony, King George Island, Antarctica. Adults take turns caring for the young. One parent incubates the eggs or guards the chicks while the other feeds or collects food.

to stand by the nest. Then, when 16–19 days old, they leave the nest to gather in groups called *crèches*, the French word for public nurseries. Once the chicks have joined the crèches the adults will lead them away across the rocks each time they come to feed them. The chicks then have to make their own way back. It is thought that this introduces them to the outside world, for soon the young will be leaving the crèche and taking to the sea themselves.

Skua and leopard seal predators

There are no land animals in the Antarctic to menace the rookeries, but predatory seabirds, especially skuas, breed near by, and take the Adélie penguins' eggs and chicks whenever the opportunity arises. Often, they wait for a penguin to neglect its eggs for a second and then swoop down. Sometimes a pair of skuas will work together, one attracting the penguin's attention while the other steals an egg. Later, the skuas wait around the crèches for a chick to become separated from its fellows. The skuas are unable to kill a healthy chick but can harass a weakened one until it succumbs.

Both adults and the young are also in danger from leopard seals as they enter the water. At one site, Prydz Bay in the Antarctic, leopard seals are estimated to take 3 percent of the adult penguin population every year.

Resource indicators

Adélies are very sensitive to human disturbance and disturbance to colonies can reduce hatching success, and possibly increases adult mortality. These and other penguins are also indicators of Antarctic resources. Increasing penguin populations are thought to be a result of either decreased competition for krill with decreasing baleen whale populations or, more likely, increased feeding or breeding success due to global warming.

AGAMA LIZARD

A male ground agama, Agama aculeata, basking in low scrub, Kalahari Gemsbok National Park, Africa. The head of this species turns bright blue during the breeding season.

AGAMAS ARE ANY OF some 60 lizard species belonging to the family Agamidae. In general agamas are about 12–18 inches (30–45 cm) in length, and most are brown or gray in color. However, the males change color during the breeding season, often turning bright shades of red, blue or yellow. The scales are usually rough to the touch, some being pointed. Some species have spiny sides to the head. Agama lizards mostly inhabit dry, semiarid or even desert environments and are found in southeastern Europe, throughout Africa and the Middle East, and through Central and south Asia to Australia.

The best-known species is the common agama, *Agama agama,* of Africa, which is 1 foot (30 cm) in length. The male of this species often has a bright terracotta head. Its body and legs are dark blue, and the tail is banded pale blue, white, orange and black. In common with other agamas, its skin is rough, like sandpaper. It has a dewlap of loose skin under its chin and a row of small spines on its neck.

Other members of the genus *Agama* include the desert agama, *A. mutabilis,* of North Africa, and the starred agama, *A. stellio,* of the eastern Mediterranean region. This last species is called the hardun or dragon lizard in some places, for example on Rhodes, in Greece. Some authorities now place this species in a separate genus, *Laukadia.* Other genera of the Agamidae family include the toad-headed agamids, *Phrynocephalus,* the spiny-tailed lizards, *Uromastix,* and the water or sail-fin lizards, *Hydrosaurus.*

Kaleidoscopic colors

The common agama or rainbow lizard is one of the most abundant reptiles in West Africa. It lives wherever the forest and bush have been cleared, including in villages and large towns. At first sight there seems to be a confusing variety of other species. They can be seen in all sizes from 5 inches (13 cm) to 1 foot (30 cm) long, some sandy, some chocolate, some with green-spotted heads and some with orange blotches on their sides. These are all, in fact, common agamas, the smallest being hatchlings, the middle-sizes females, and the largest of all, males. The bright orange-and-blue coloring occurs only in the breeding season, and even then only in dominant males. Subordinate males are dull brown in color.

During the day the common agama is extremely active, running across open spaces from one heap of stones to another, darting out to snap up ants, even leaping into the air after flying insects. If caught in the open, it is able to run on its hind legs. Only in the afternoon, when the temperature reaches around 100° F (38° C) in the shade, does it try to find a cool place in which to rest. Toward dusk, it congregates in communal roosts, often in the eaves of houses, and at night all the males, including the dominant ones, go a dull brown color all over. The next morning, when out in the early sun, their brilliant colors return.

Temperature control

The desert agama lives in the dry areas of North Africa and is distinctive for the ring of spiky scales around its tail. It is commonly found in northern Egypt. After the cold night the agama will be literally stiff with cold, but as the sun rises it absorbs heat and its temperature increases so that it is able to resume feeding and moving about. As the sun's power increases, the desert agama must be careful not to overheat, although it can tolerate high temperatures. The sparse scrub of the desert usually gives it sufficient shelter as it dashes from one bush to another.

AGAMA LIZARDS

CLASS	**Reptilia**
ORDER	**Squamata**
SUBORDER	**Sauria**
FAMILY	**Agamidae**
GENUS	***Agama* and others**
SPECIES	**About 60, including common agama, *Agama agama*; desert agama, *A. mutabilis*; and starred agama, *A. stellio***

ALTERNATIVE NAMES
**Common agama: rainbow lizard.
Starred agama: hardun; dragon lizard.**

LENGTH
Normally 12–18 in. (30–45 cm)

DISTINCTIVE FEATURES
Generally brown or gray, but males undergo marked color change during mating season, becoming bright red, blue or yellow; scales usually rough to the touch; pointed, spiny scales at sides of head (some species only)

DIET
Mainly large insects; also grasses, berries, seeds and eggs of other lizards

BREEDING
Varies between species. Breeding season: summer months (subtropical climates), early part of dry season (Tropics); number of eggs: 2 to 20 per clutch; breeding interval: several clutches per year.

LIFE SPAN
Probably 10–15 years

HABITAT
Mainly dry, semiarid or desert environments

DISTRIBUTION
Southeastern Europe; throughout Africa; Madagascar; from Middle East across Central and southern Asia; Australia

STATUS
Some species abundant; none threatened

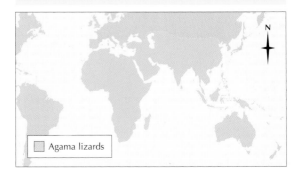

Agama lizards

Prey on insects

Agamas are mainly insectivorous, chasing their prey at speed and catching small insects with the tongue or snapping up larger ones directly with the mouth. The incisorlike front teeth are pointed like those of insectivorous mammals. Agamas sometimes also eat grasses, berries, seeds and the eggs of smaller lizards.

Polygamous breeders

The common agama is polygamous and the brightly colored male may be seen with half a dozen or more females. He maintains a territory, which he defends vigorously. In rural districts such territories are well spaced out, each with a landmark such as a tree, log or rock, near the middle. In towns there are often higher densities of agamas competing for space, and this can lead to more frequent fighting between males.

In courtship the male comes alongside the female, bobbing his head. If she is in breeding condition, she allows him to grip her neck with his jaws. He then puts one hind leg over her back, and twists the rear end of his body under her. The female then raises her tail away from the male and the vents are brought together. Sometimes the female initiates the courtship by running up to the male and raising her tail in front of him.

Agamas have a very definite breeding season. In the common agama this occurs after the "long rains" of March to May. The males have fertile spermatozoa all the year, but the females can lay eggs only from June to September, some months after the rains. At this time the vegetation becomes lush and insect numbers rise, providing the female agamas with an ample supply of protein for the formation of eggs. These are laid in clutches of between 2 and 20, depending on the species.

This male common agama is in full breeding condition. Polygamous breeders, one male agama will defend a territory containing perhaps six or more females.

AGOUTI

AGOUTIS ARE A GENUS of Central and South American rodents and resemble large, long-legged guinea pigs or cavies. The numerous species vary in color from tawny to blackish brown with lighter underparts. Some forms have white stripes. Their coarse hair is longer on the hindquarters, where it is usually bright orange or golden but may be white or black. This hair is raised when the animals are alarmed or being aggressive. Their heads are rat-like in appearance, with relatively large, pinkish ears. Agoutis are about 1⅓ feet (40 cm) long, with a short, hairless tail and long legs. They have five toes on the forefeet and three on the hind feet, all of which have hooflike claws.

A close relative of the agoutis is the acouchi, *Myoprocta pratti*, of northwestern South America. The principal difference is that the acouchi is smaller and has a slender, white-tipped tail that is used as a signal in courting ceremonies.

Forced into nocturnal living

Agoutis are abundant in forest, savanna, bush and wooded areas throughout Central and South America from Mexico southward to Brazil and Peru. They are sometimes found on farmland and have been introduced to the Caribbean islands. Agoutis are basically diurnal (day living) but have become crepuscular, moving about in the twilight hours, and are nocturnal (night living) where they have been disturbed by people.

Central American agouti, Dasyprocta punctata, *one of 11 species of agoutis. Many of these shy animals have been forced into living mainly by night because of human disturbance.*

Where this is the case they spend the day in holes in trees or in burrows scraped in the ground among soft limestone boulders or under tree roots.

The burrows are shallow "foxholes," 2–3 feet (60–90 cm) deep, sometimes roofed over by a lattice of twigs covered with leaves. Each burrow is occupied by one animal or a small family group. Well-worn tracks radiate from the entrance to the communal feeding grounds.

Reports differ as to whether agoutis are social or solitary in their way of life. Different species may well have different habits. However, the basic social unit is one mating pair, which stays together for life. It seems that, although they sleep in their burrows alone or in small groups, agoutis gather in groups of up to 100 animals to feed. They are very shy but have also been described as being "highly strung," fighting fiercely among themselves yet fleeing in panic at the first hint of danger. Agoutis have been known to live up to 17 years in captivity, but the average length of life is thought to be around 6 years.

Delicate eaters

Like all rodents agoutis are mainly vegetarians, browsing on leaves, fallen fruits and roots, occasionally climbing trees to take unripe fruits. Sometimes agoutis damage commercial crops such as banana plants and sugarcane. Their habit of eating fallen fruits means that they act as seed dispersers by carrying and burying fruits. They have also been known to take the eggs of ground-nesting birds, and might search for shellfish on the seashore.

Agoutis are delicate eaters, sitting back on their haunches and holding their food in the forefeet and, if it has a tough skin, peeling it carefully with their teeth before eating it. They hoard food in small stores buried near landmarks.

Advanced young

Litters number from one to four, although one or two young is most common. Some species have two litters each year, in May and October, others appear to breed all the year round. The young are born in a burrow lined with leaves, roots and hair. Their arrival is unusual in that the mother gives birth while in a squatting position. The young are precocial, meaning that they are quite well developed at birth and capable of a high degree of independent activity. They are covered with

AGOUTIS

CLASS	**Mammalia**
ORDER	**Rodentia**
FAMILY	**Dasyproctidae**
GENUS	***Dasyprocta***
SPECIES	**11, including Central American agouti, *D. punctata***

WEIGHT
2–8 lb. (1–3.6 kg)

LENGTH
Head and body: 1⅓–2 ft. (40–60 cm); tail: ½–1½ in. (1.3–3.8 cm)

DISTINCTIVE FEATURES
Resemble large, long-legged guinea pigs or cavies; large nose; ratlike ears; long legs with elongated hind feet; hooflike claws; short, hairless tail; color varies from tawny to blackish brown with lighter underparts; hair is coarse and longer on hindquarters

DIET
Mainly fallen fruits but also leaves and roots; occasionally crops such as sugarcane and banana plants

BREEDING
Age at first breeding: not known; breeding season: varies with species, often all year; number of young: usually 1 or 2; gestation period: 90–120 days; breeding interval: 1 year (most species), 2 litters per year (other species)

LIFE SPAN
Up to 17 years in captivity

HABITAT
Forest, bush, savanna and farmland

DISTRIBUTION
Central and northern South America

STATUS
Most species common; endangered: 2 species; vulnerable: 1 species

hair and are born with their eyes open. Within an hour they are nibbling at vegetation. This advanced stage of development is linked with the long gestation period of 90–120 days.

While the young are very small the father is barred from the nest. They remain with the parents for some weeks.

A black agouti,
Dasyprocta fuliginosa,
of the upland Amazon
rain forest, Amazonas
State, Brazil.

Wary animals
On being disturbed an agouti first freezes to avoid being detected, sitting with its body upright and ankles flat on the ground ready to leap off at speed. The young will also behave in this way from birth. If its alarm continues, the agouti will scream shrilly and flee, dodging obstacles with remarkable agility.

Agoutis have been described as the "basic diet of South American carnivores," such as the ocelot and the jaguar. Humans also find in agoutis a plentiful source of food, and their flesh is often prized by local people. Agoutis are hunted, too, in areas where they are pests of sugarcane and banana plantations.

Adapted for running
Agoutis are fast runners, escaping predators by speed rather than by hiding in burrows. They are agile and bound through undergrowth undaunted by precipices, on which they display the agility of goats. Leaps of 20 feet (6 m) from a standing start have been recorded. The reason for their speed and agility lies in the anatomical features that distinguish the agoutis from their relatives, namely their long, thin legs and the hooflike claws on which they walk and gallop. In these they resemble the ungulates or hoofed mammals, these last having become adapted for running by the development of long legs, the reduction in the number of toes and the formation of hooves. There has been a similar trend in the agoutis.

ALASKA BLACKFISH

The Alaska blackfish is able to withstand extreme conditions. Legend has it that it will even survive if frozen into a block of ice.

A FISH THAT LIVES IN weed-choked swamps and ponds in lowland areas of Alaska and the Chukotsk Peninsula, Siberia, the Alaska blackfish is seldom found far inland. In the past it was important commercially in some areas, and served as a food source for the native Inuit people and their dogs. Such an extensive fishery is no longer in existence, although it is still used to some extent for human food.

The species itself is somberly colored, adults being dark brown above, with four to six irregular black bars on the sides. The underside is pale with dark brown sparkling. The fish reaches a maximum length of about 8 inches (20 cm).

Freshwater dweller

In winter the Alaska blackfish lives in deep water, perhaps down to 20 feet (6 m), moving back into depths of a few inches in the spring. In the summer it lives among dense growths of water plants and never enters clear water. It is not an active fish, although it can move at lightning speed when alarmed or hunting prey.

The blackfish is specialized for surviving adverse conditions. In winter it lives below the ice at temperatures approaching 32° F (0° C) and can exist in water with a low oxygen content. It is also tolerant of overcrowding and can stand competition with other species as well, both for living space and food.

Diet and breeding

The blackfish feeds by slowly sculling (using its pectoral fins) to within inches of its prey and then making a sudden dart forward. Its main food is insect larvae, especially those of the two-winged flies, such as midges and mosquitoes, along with crustaceans, ostracods (water fleas) and snails.

The blackfish spawns in July, and in May and June the males develop a reddish margin to the fins. There is no parental care of the eggs.

CLASS	**Osteichthyes**
ORDER	**Esociformes**
FAMILY	**Umbridae**
GENUS AND SPECIES	***Dallia pectoralis***

ALTERNATIVE NAMES
Alaska fish; devilfish

WEIGHT
1¾–3 oz. (50–60 g)

LENGTH
Up to 8 in. (20 cm)

DISTINCTIVE FEATURES
Backward placement of dorsal fin; 3 pelvic fins; well rounded pectoral and tail fins. Adult: dark brown above with 4 to 6 bars on sides; pale below with dark brown sparkles; fins speckled and dorsal, anal and tail fins rimmed with white (reddish in male at spawning). Juvenile: lighter brown; dorsal and anal fins shorter in female than male.

DIET
Insect larvae, crustaceans, ostracods (a type of crustacean or water flea) and various snails

BREEDING
Age at first breeding: 2–3 years, when about 2 in. (5 cm) long; breeding season: July; number of eggs: 100 to 300; hatching period: about 9 days under good conditions

LIFE SPAN
Up to 8 years

HABITAT
Typically weed-choked, lowland swamps and ponds; also streams, rivers and large lakes if there is abundant vegetation

DISTRIBUTION
Chukotsk Peninsula, Siberia, and lowland areas of westernmost Alaska

STATUS
Not known

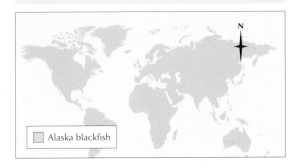

Alaska blackfish

ALBATROSS

THE LARGEST FAMILY OF birds in the petrel order, and the largest of all flying birds, albatrosses have a goose-sized body with very long, slender wings. Of the 14 species, one of the largest is the wandering albatross, *Diomedea exulans*, which has a wingspan that can exceed 11 feet (3 m). The plumage is black and white or, in a few species, brown. In only some of the species is it possible to tell the sexes apart.

Ocean wanderers

Nine species of albatrosses are confined to the Southern Hemisphere, breeding mainly on the subantarctic and oceanic islands. Another three are found in the North Pacific, with the waved albatross, *Diomedea irrorata*, on the equatorial Galapagos Islands. None breed in the North Atlantic, although fossil remains have been found in England and a few have been recorded as vagrants in modern times. These vagrants include wandering, black-browed, yellow-nosed, gray-headed and light-mantled sooty albatrosses. In 1860, one black-browed albatross appeared in a gannet colony south of Iceland on the Faroe Islands. It accompanied the gannets on their annual migrations for 30 years, until it was shot.

The doldrums, the windless belt around the equator, are possibly one of the reasons why so few albatrosses have been recorded in the North Atlantic, as albatrosses need a sustained wind for flight. They are heavy birds with comparatively small wing muscles, but they can remain airborne for long periods and cover vast distances. This is because of the difference between the wind speed at the water's surface, where it is slowed down by friction, and some 50 feet (15 m) above. The albatross glides swiftly downwind toward the water's surface, gathering speed. When it is just above the water it swings sharply round into the wind and soars up. As it rises it loses momentum, causing its ground speed, in relation to the water surface, to decrease. Its air speed, however, does not decrease so fast, because the bird is continually meeting faster wind currents as it rises. By the time the air speed has dropped completely the albatross will have gained sufficient height to start the downward glide again.

The main haunt of the albatrosses is the subantarctic zone. The strong winds at these latitudes sweep around the world and there is nearly always enough wind to keep the albatrosses aloft, although they can also glide in quite gentle breezes. To increase its speed, an albatross partly closes its wings, which reduces the air resistance without seriously affecting its lift. However, with their great wingspan and weak wing muscles, albatrosses have difficulty in taking

A wandering albatross shows off its enormous wingspan, which can be as much as three times the length of its body, as it glides over the Antarctic waters.

A black-browed albatross watches over its only offspring.

BLACK-BROWED ALBATROSS

CLASS	**Aves**
ORDER	**Procellariiformes**
FAMILY	**Diomedeidae**
GENUS AND SPECIES	***Diomedea melanophris***

WEIGHT
About 5 lb. (2.3 kg)

LENGTH
Head to tail: 2½–3 ft. (75–90 cm); wingspan: 6⅔–8 ft. (2.15–2.45 m)

DISTINCTIVE FEATURES
Extremely long, slender wings; strong bill with hooked tip and large nostril tubes; webbed feet. Adult: white head and underparts; straw-yellow bill; dark wings and back with distinct white stripe on underwing. Juvenile: gray plumage.

DIET
Squid, fish, shrimps and mollusks; occasionally smaller seabirds

BREEDING
Age at first breeding: 7 years; breeding season: October onwards; number of eggs: 1; incubation period: 70–80 days; fledging period: 116 days; breeding interval: 1 year

LIFE SPAN
Up to 34 years

HABITAT
Seas and oceans, from edge of Antartic ice to temperate and subtropical waters

DISTRIBUTION
Breeds throughout subantarctic zone on oceanic islands; travels vast distances when not breeding, often north to subtropics

STATUS
Fairly common

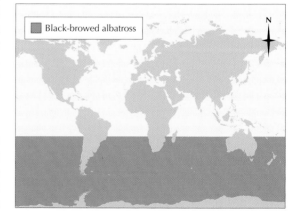

Black-browed albatross

off. When there is enough wind, especially if there are thermal currents or eddies, takeoff is easy. However, on still days they have to taxi, running along and flapping their wings until they have gained sufficient air speed. Otherwise, they drop over the cliff face and glide away.

Some species are fairly confined in their range, such as Buller's albatross, *Diomedea bulleri*, in New Zealand. Others, like the wandering, black-browed and sooty albatrosses, circle the world from the Tropics to Antarctica.

Marine feeders

All species of albatrosses feed on marine organisms living near the surface of the sea, such as fish, squid and crustaceans. They also take small seabirds on occasions, and they like to feed on refuse from ships, flopping down into the water as soon as a bucketful is tipped overboard. Legend has it that sailors who have fallen overboard have been attacked by albatrosses.

Clifftop breeding sites

Breeding grounds, where albatrosses gather in tens of thousands, are usually on the top of cliffs where the birds can take off easily. Albatrosses are extremely faithful to their nest sites, and populations have survived such calamities as volcanic eruptions or pillage by humans because the immature birds that were absent at the time later returned to breed.

Albatrosses are very long-lived birds. A wandering albatross may live for 70 years. They do not start breeding until at least 7 years old, but young birds return to the breeding ground before then and court halfheartedly. The courtship displays of wandering albatrosses, which can be seen throughout the breeding season, are most spectacular. The two birds of a pair dance around awkwardly with outstretched wings, trumpeting and snapping.

A single egg is laid in a cup-shaped nest of mud and plants and is incubated by both parents for periods ranging from 65 days in the smaller species to 81 days in the larger ones. The chick is also brooded for a short time and is guarded by the adults for several weeks. It is then left by itself and both parents can be away feeding at once, returning at intervals to give the chick a huge meal of regurgitated squid, crustaceans or fish. The young of the smaller albatrosses fledge in 2 to 3 months, but larger ones may spend 8 or 9 months in the colony, sitting out the severe southern winter until summer arrives. The parents feed them the whole time, so by the time the young become independent it is too late for the parents to nest again and breeding is only possible in alternate years.

The young albatrosses leave the breeding grounds to glide away around the oceans. Before they return to court several years later, they may circle the globe many times.

No natural predators

Living as they do on remote islands, albatrosses have no natural predators except for skuas and giant petrels that occasionally take eggs and young. Any introduced carnivores would, however, wreak havoc among the densely packed nests, for the sitting albatross's reaction to disturbance is just to sit tight on the nest and clack its bill. It also spits oil from digested crustaceans and fish, but this would not discourage a determined attacker.

The sailors' curse

Albatrosses have been known to sailors since the days of Magellan. They were considered to be harbingers of wind and storms, which is not surprising in view of their difficulty in remaining aloft in calm weather. They were also regarded as the reincarnations of seamen washed overboard. It was thought extremely unlucky to kill an albatross, as Coleridge wrote in his famous poem, *The Rime of the Ancient Mariner*. But, despite the ancient mariner's punishment of having an albatross hung around his neck and

A pair of albatrosses engage in a ceremonial grooming session at their nest site on Campbell Island, south of New Zealand.

suffering the tragic misfortunes that later befell him, sailors have not always treated albatrosses kindly. Their capture on baited hooks trailed from the stern of a ship often relieved the monotony of the sailors' diet.

Threats to albatrosses

Not only were albatrosses caught to provide a luxury meal for sailors, the 19th century produced an appetite for birds' feathers. These became a favorite material in the hat-making trade. The wings were sometimes cut off still-living birds. Luckily, however, this fashion petered out before all the birds were dead.

World War II introduced another crisis for albatrosses. Long-range aircraft flights made it necessary for many oceanic islands, such as Midway Island, home of the Laysan albatross, *Dionedea innutabilis*, to be turned into military posts. Not only do albatrosses use the United States Navy's runways for taking off, they also soar in the thermals above them, providing a serious danger to aircraft. Of the many methods people have employed to reduce this danger, the most effective has been the bulldozing of dunes by the runways, which cause the updraughts that the albatrosses need for flying.

A more recent threat to albatrosses has appeared with the worldwide development of longline fishing. The vessels trail lines up to 80 miles (130 km) long, with up to 12,000 baited hooks on them. Just one fishery can be responsible for the deaths of 145,000 seabirds per year.

ALLIGATOR

THERE ARE TWO SPECIES of alligators, reptiles which, with the caimans, belong to a family closely related to the crocodiles. Alligators and crocodiles look extremely alike.

The main distinguishing feature is the teeth. In a crocodile the teeth in its upper and lower jaws are in line, but in an alligator, when its mouth is shut, the upper teeth lie outside the lower ones. In both animals the fourth lower tooth on each side is perceptibly larger than the rest. In the crocodile this tooth fits into a notch in the upper jaw and is visible when the mouth is closed, whereas in the alligator, with the lower teeth inside the upper, it fits into a pit in the upper jaw and is lost from sight when the mouth is shut. In addition, the alligator's head is broader and shorter, and the snout consequently blunter, than in the crocodile. Otherwise, especially in their adaptations to an aquatic life, alligators and crocodiles have much in common.

It is sheer accident that two such similar reptiles should have been given such different common names. The reason is that when the Spanish seamen, who had presumably no knowledge of crocodiles, first saw large reptiles in the Central American rivers, they spoke of them as lizards, *el largato* in Spanish. The English sailors who followed later adopted the Spanish name but ran the two into one to make "allagarter," which was later further corrupted to "alligator."

Smaller than they were

One of the two species of alligators is found in North America, the other in China. The Chinese alligator, *Alligator sinensis*, averages a little over 4 feet (1.2 m) in length. The American alligator, *A. mississippiensis*, is much larger, with a maximum recorded length of 20 feet (6 m). This length, however, is seldom attained nowadays because the American alligator is legally harvested for its skin. Whenever there is intense persecution or hunting of an animal, the larger ones are quickly eliminated and the average size of the remaining population drops slowly. However, adult male American alligators of 13 feet (4 m) in total length are still fairly common.

An American alligator eating a raccoon in the Florida Everglades. Mature adults feed mainly on fish, but they will also capture small mammals that come to drink at the water's edge.

AMERICAN ALLIGATOR

CLASS **Reptilia**

ORDER **Crocodylia**

SUBFAMILY **Alligatoridae**

GENUS AND SPECIES *Alligator mississippiensis*

LENGTH
Adult male: normally less than 13 ft. (4 m); occasionally up to 16⅔–20 ft. (5–6 m)

DISTINCTIVE FEATURES
Adult: bony nasal bridge; large, broad, robust skull; no osteoderms (bony deposits) in belly scales; upper teeth lie outside lower teeth when mouth shut. Juvenile: black skin with 4 or 5 yellow crossbands on body and 10 or 11 bands on tail; crossbands fade with age.

DIET
Adult: other reptiles, fish, small mammals and birds; also large mammals (very large crocodiles only). Juvenile: insects, mollusks and freshwater shrimps.

BREEDING
Age at first breeding: 6 years; breeding season: late April–early June; number of eggs: around 40; hatching period: 62–66 days; breeding interval: 1 year

LIFE SPAN
Probably up to 50 years

HABITAT
All available aquatic habitats including swamps, rivers, lakes, tidal zones and ponds

DISTRIBUTION
Southeastern U.S. from Virginia border and North Carolina south to southern Florida; range extends west to Rio Grande, Texas, and northwest to southern Arkansas and McCurtain County, Oklahoma

STATUS
Locally common; estimated population: 800,000 to 1,000,000

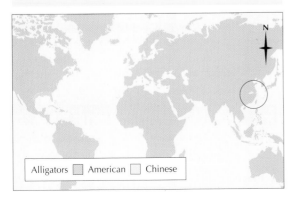

Alligators ▩ American ☐ Chinese

Long lazy life

Alligators are more sluggish than crocodiles and this possibly affects their longevity. They spend most of their time basking in swamps and on the banks of lakes and rivers. There are records of alligators having lived for more than 50 years.

The American alligator is restricted to the southeastern United States from the Virginia border and North Carolina to southern Florida. Its range then extends west to the Rio Grande in Texas and northwest as far as southern Arkansas and McCurtain County, Oklahoma. The Chinese alligator, meanwhile, is found only in the Yangtze River Basin in China.

Alligators' food changes with age. The juveniles feed on insects, mollusks and freshwater shrimps. As they grow older they take more frogs, snakes and fish. Mature adults live mainly on fish but will catch muskrats and small mammals that go down to the water's edge to drink. They also take a certain amount of waterfowl. Very large alligators may occasionally pull large mammals such as deer or cows down into the water and drown them, and will also attack humans.

Alligator nests

The female alligator plays the more active role in courtship and territorial defense. The males spend much of the breeding season quarreling among themselves, roaring and fighting. The roaring attracts the females to the males, as does a musky secretion from glands in the male's throat

Female Chinese alligator, eastern China. In addition to persecution by humans, this species is suffering from habitat destruction. Very few individuals remain in the wild.

and cloaca. Courtship usually takes place at night and is slow and protracted. The pair bump against each other, pressing on the head and neck. The male pushes the female under the water before mounting, with his lower abdomen curled under hers for a couple of minutes.

A large nest-mound is made for the reception of the eggs. The female scoops up mud in her jaws and mixes it with decaying vegetation. The mixture is then deposited on the nest site until a mound 3 feet (90 cm) high is made. The eggs are hard-shelled and number 40 on average. They are laid in a depression in the top of the mound and covered with more vegetation. The female remains by the eggs until they hatch 62–66 days later, incubated by the heat of the nest's rotting vegetation.

Sex determination

Whether the young hatch as males or females is decided by temperature, as in many reptiles. Different parts of the mound experience different temperatures due to exposure to the sun and so eggs placed in different parts of the mound will develop into different sexes. This system leaves the possibility that the sex ratio may be determined by the mother, although this is not proven.

The hatchling alligators peep loudly and the female removes the layer of vegetation over the nest to help them escape. Baby American alligators are 8 inches (20 cm) long when first hatched

Male alligators fighting in the breeding season, St. Augustine, Florida.

and grow approximately 1 foot (30 cm) a year, reaching maturity at 6 years. However, it may be several years before they breed for the first time.

Threats to survival

Young alligators are taken by predatory fish, birds and mammals, and at all stages of growth they are attacked and eaten by larger alligators. This natural predation was, in the past, just sufficient to keep their numbers steady. Then came the fashion for making shoes, handbags and other ornamental goods of alligator skin. This was the driving force behind the decrease in alligator numbers, as well as those of other crocodilians worldwide, in the early part of the 20th century. However, with careful, sustainable harvesting of wild populations, legally conducted throughout the entire range of the American alligator, numbers have increased greatly. The continued survival of this species has probably been ensured, although land drainage and pollution are now the main threats to alligators.

The Chinese alligator is a far more serious case. Its flesh is eaten and many parts of its body are used as charms and aphrodisiacs, and for their supposed medicinal properties. In addition, the recent further damming of the Yangtze River continues to threaten the future of the Chinese alligator. Few individuals are thought to remain in the wild. There are captive breeding programs in both China and the United States.

Pets down the drain

There was for a while another commercial interest detrimental to the alligator: while the fashion for skins from larger individuals was at its height, a fashion for alligator pets set in. Baby alligators were netted in large numbers for pet shops.

This fashion had its disadvantages for owners as well as the alligators. Obviously, the alligator achieves much too large a size for it to be convenient in a modern house or apartment. Most people who invested in an alligator found it necessary to dispose of the animal soon after, and zoos proved unable to deal with the quantity offered them. Brookfield Zoo near Chicago, for example, built up an enormous herd from unwanted pets. It is often said that alligators are disposed of in such a way that they end up in the sewers. Headlines have appeared in the press to the effect that the sewers of New York are teeming with alligators that prey on the rats and terrorize sewer workers. However, such reports are no doubt exaggerated.

AMOEBA

MOEBAE FORM A group of the single-celled organisms called Protozoa, in the kingdom Protista. Like any cell, an amoeba basically consists of an envelope containing protoplasm (see below). In the middle of the cell, surrounded by the protoplasm, is the nucleus, a body that can be thought of as a control center for the organization of the cell's activities.

The protoplasm is not, as was once thought, a jelly. It actually has a very complicated structure. It consists of a cytoplasm divided into a granular endoplasm and, at the ends of the pseudopodia (footlike protrusions) and elsewhere under the surface, a clearer layer known as ectoplasm.

Many amoebae

The name amoeba is applied not only to members of the genus *Amoeba* but also to a range of different types of protozoans with pseudopodia living in the sea, in fresh water, in damp soil and in the bodies of larger animals. They include some organisms with shells, such as *Arcella*, and also the half-dozen or so species that live in the human mouth and digestive system, one of which, *Entamoeba*, is the cause of amoebic dysentery. Some amoebae contain many nuclei, rather than just one, among them the giant *Chaos carolinensis*, which may measure up to ⅕ inch (5 mm) in diameter.

Amoeba proteus, the textbook amoeba, measures about ¹⁄₅₀ inch (0.5 mm), is just visible to the naked eye and may be found in fairly still, fresh water. It moves by extending a finger of protoplasm called a pseudopodium, or false foot. As the pseudopodium enlarges, the cell contents (the protoplasm and nucleus) flow into it, while the rest of the cell contracts behind. Although it has no definite shape, an amoeba is not a shapeless sac of protoplasm. It has a permanent hind end and forms its pseudopodium in a characteristic pattern according to the species.

Feeding

The amoeba feeds on other small protozoans. It captures its prey by flowing around it, the protoplasm completely surrounding the food to enclose it in a "food vacuole," containing fluid in which the prey had been swimming. Digestion is a similar process to that occurring in many other

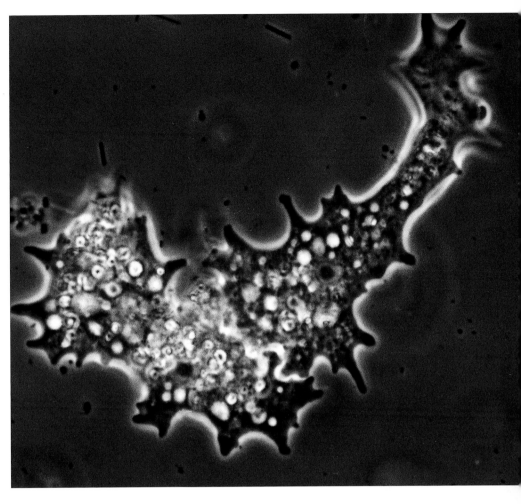

organisms. The amoeba secretes digestive juices into the food vacuole and the digestible parts of the protozoan are broken down and absorbed. The undigested remains are left behind as the amoeba moves along. This process is known as phagocytosis. In a similar process, called pinocytosis, channels are formed from the cell surface, leading into the cell. The amoeba draws fluid into the channels and from their tips vacuoles are pinched off. It then absorbs fluid into the protoplasm in the same way as the digested contents of the food vacuoles. This is a method of absorbing fluids in bulk into the cell.

Water is continually passing in through the cell membrane as well as being brought in by phagocytosis and pinocytosis. Excess fluid is pumped out by contractile vacuoles that fill with water and then collapse, discharging the water to the outside.

Reproduction

Different amoebae have different reproductive strategies. *Amoeba proteus*, for example, reproduces itself by dividing into two equal parts, a

A pair of amoebae in the process of reproducing sexually. Many fingerlike pseudopodia, or false feet, can be seen on the amoebae in this photograph.

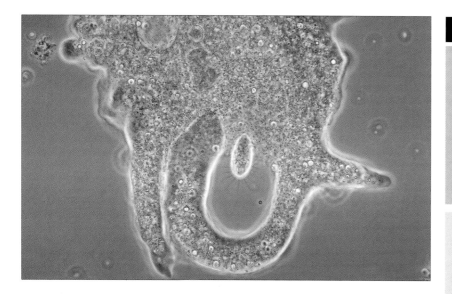

An amoeba, Amoeba discoides, *engulfs a small food particle (center of photograph).*

AMOEBAE

KINGDOM	**Protista**
PHYLUM	**Sarcomastigophora**
SUBPHYLUM	**Sarcodina**
CLASS	**Lobosa**
ORDER	**Amoebida**
GENUS AND SPECIES	**Hundreds**

LENGTH
**Chaos carolinensis: up to ⅕ inch (5 mm);
Amoeba proteus: ¹⁄₅₀ inch (0.5 mm)**

DISTINCTIVE FEATURES
Single-celled organisms; form pseudopodia (temporary, footlike protrusions, or false feet)

DIET
Microscopic particles

BREEDING
Many different strategies such as binary fission (dividing in two) and multiple fission (dividing into many cells); some species reproduce sexually

LIFE SPAN
Varies according to environment

HABITAT
Seas, stagnant fresh water and damp soil; also live parasitically on a variety of other, larger animals

DISTRIBUTION
Worldwide

STATUS
Superabundant

process known as binary fission. This process takes less than an hour. It begins with the amoeba becoming spherical. The nucleus then divides into two, and the two halves move apart. The cell then splits down the middle.

Other species reproduce in a different manner. Multiple fission, in which multinucleated amoebae (amoebae with many nuclei) divide into many smaller ones, is also common. This might lead to spore formation, in which each nucleus becomes surrounded by a little cytoplasm and a tough wall, all within the original cell. The resulting "cysts" can survive if the water dries up and can be dispersed to found new populations. Larger cysts may be formed without reproduction taking place, when the whole cell surrounds itself with a thick wall. Though some amoebae reproduce sexually, *Amoeba proteus* has never been seen to do so.

In certain shelled amoebae, the parent sporulates (forms many spores), producing a large number of small, naked amoebae, each of which produces its own shell. Finally, hologamy, the fusion of two individuals, is seen in some species.

Pushing or pulling?

Years ago scientists could watch amoebae only from above in the usual manner of looking at very small objects. From this angle one could see the pseudopodia advancing over the surface of the microscope slide and apparently in contact with it. More recently a technique was devised for watching amoebae from the side and a new detail has come to light. In fact, when each pseudopodium moves forward it is supported by an extremely small peg of protoplasm that remains attached to the ground at one spot while the rest of the animal, raised just above the ground, advances over it. As the movement is completed, the pseudopodium is withdrawn and reincorporated into the body of the amoeba.

A number of theories of amoeboid movement have been proposed. One can see, under the higher powers of the microscope, the protoplasm streaming forward along the center of the pseudopodium and moving out to the sides at the tip in what has been descriptively named the fountain zone. It then acquires a firmer consistency. At the same time the reverse change occurs at the tail, where the protoplasm resumes its forward flow.

What is still in doubt is whether the advancing protoplasm is being pushed forward from behind, or pulled by changes in the proteins in the fountain zone. The problem is by no means trivial, for some of our own cells move in an amoeboid manner; its solution in terms of the behavior of protein molecules could cast light on one of the basic properties of protoplasm.

ANACONDA

THE LARGEST SNAKES ARE to be found in the boa family, and the largest of these is *Eunectes murinus*, the anaconda or water boa. Large though it is, few animals have been the subject of such exaggeration in respect of their size. Claims for 140-foot (43-m) anacondas have been made in the past and the figure 40 feet (12 m) often occurs in travel literature. In fact anacondas seldom exceed 24½ feet (7.5 m) and most individuals are much smaller than this.

The anaconda is olive green with large, round black spots along the length of its body and two light, longitudinal stripes on its head. It lives throughout tropical South America, east of the Andes, mainly in the Amazon and Orinoco River Basins, and in the Guianas. Its range extends north to Trinidad. The species is variable in color and size, giving rise to numerous sub-specific names. However, these can be regarded as merely geographical variations. The closely related *Eunectes notaeus* of Paraguay is known as the Paraguayan or southern anaconda.

Life in streams and swamps

Water boa is a good alternative name for the anaconda, the most aquatic of the boas. It is apparently never found far from water, sluggish or still water being preferred to rapid streams. It is this preference that limits the species to the river basins east of the Andes. Swamps are a favorite haunt. Anacondas have, as a rule, fixed hunting grounds and generally live alone, but they are occasionally seen in groups.

Largely nocturnal in habit, anacondas lie up during the day in the shallows or sunbathe on low branches, usually over water. On land they are relatively slow moving, but they are able to swim rapidly. However, they often float motionless in the water, allowing the current to carry them downstream.

Killing by constriction

Anacondas usually lie in wait for their prey to come down to the water's edge to drink. They strike quickly with the head, grabbing the animal and dragging it underwater so that it drowns. At other times anacondas may actively hunt prey on land. They usually feed on birds and small mammals, such as deer and peccaries, and on large rodents, such as agoutis. Fish form a large part of their diet, and turtles and caimans are also sometimes attacked.

The method of killing the prey is the same as in other constricting snakes, such as the pythons. The prey is not crushed, as is sometimes supposed, but merely contained. Each time the victim exhales, the coils of the anaconda tighten so that the ribs cannot expand, until it suffocates.

Anaconda constricting a caiman. Each time its victim exhales, the snake tightens its coils until the animal suffocates. Alternatively, the anaconda drags prey under the water, drowning it.

Anacondas often feature in folklore and tales of giant, man-eating snakes. However, the largest specimens seldom grow longer than 20 feet (6 m), and there are few confirmed instances of their attacking humans.

The special jaw attachment that snakes have allows the anaconda to swallow victims larger than itself. The upper and lower jaws are only loosely attached, and the brain is protected from pressure by massive bones. Also a valve on the breathing tube allows the snake to breathe while swallowing. After a large meal, which might satisfy an anaconda for several weeks, the snake rests for a week or more until digestion has taken place. However, its diet will usually consist of more frequent, smaller meals.

Breeding

Few observations have been made on the breeding cycle of the anaconda. Males of southern anacondas studied in captivity were apparently aroused by the scent of the females. In this species the male moves up alongside the female, flicking his tongue over her, until his head is resting over her neck. When in this position, he erects his spurs, two clawlike projections, and moves them against the female's skin. When the cloacal regions are in opposition, a hemipenis is inserted and mating takes place.

Anacondas, like other boas, are ovoviviparous and give birth to live young. From 20 to 40, sometimes more, young are born, usually in February or March. Each baby is 2–3 feet (60–90 cm) long.

Anacondas in folklore

South American native peoples have numerous stories about the anaconda, from the belief that it turns itself into a boat with white sails at night, to the mythology of the Taruma Indians, who claimed to be descended from an anaconda.

Along with stories of venomous qualities and body size, there is also exaggeration about the danger involved in meeting an anaconda. There are, however, remarkably few, if any, authentic stories of people killed and eaten by this snake.

ANACONDA

CLASS	**Reptilia**
ORDER	**Squamata**
SUBORDER	**Serpentes**
FAMILY	**Boidae**
GENUS AND SPECIES	***Eunectes murinus***

ALTERNATIVE NAME
Water boa

LENGTH
Up to 24½ ft. (7.5 m), but usually much smaller than this

DISTINCTIVE FEATURES
Very large size; varies in color but often olive green with large, round, black spots along length of body; often has 2 light, longitudinal stripes on head

DIET
Fish and mammals, especially small deer, peccaries and large rodents such as agoutis; also birds, turtles and caimans

BREEDING
Little known. Ovoviviparous, giving birth to live young; number of young: usually 20 to 40, born in February or March.

LIFE SPAN
Probably 40–50 years

HABITAT
Forested habitats near rivers and swamps; prefers slow-moving or still water

DISTRIBUTION
Tropical South America east of Andes Mountains: Amazon and Orinoco River Basins and in the Guianas; range includes island of Trinidad

STATUS
Uncommon; rare in some areas due to persecution

Anaconda

ANCHOVY

THE ANCHOVY FAMILY comprises some 139 species, similar to herrings but smaller. Their maximum size, attained by the New Guinea thryssa, *Thryssa scratchleyi*, is 14½ inches (37 cm). The anchovy familiar to Europeans is the European anchovy, *Engraulis encrasicolus*, which is canned or converted to anchovy paste. It is about 8 inches (20 cm) in length and found in the Eastern Atlantic, from Norway south to South Africa. It is also found in the Mediterranean, Black Sea and western Indian Ocean. The anchovy familiar to Americans is a similar fish, the northern or California anchovy, *E. mordax*. A little larger than the European anchovy, it ranges from northern Vancouver Island, British Columbia, south to Cape San Lucas, Baja California, and is used in quantities by fishers as live-bait. Anchovies elsewhere in the world are also fished commercially.

Anchovies have a rounded, protuberant snout and a very long jaw. Their bodies are slender, translucent, clear green on the back, bright silvery on the sides and silver white below.

Vast shoals

Anchovies are most abundant in tropical seas, but large numbers also live in the shallow parts of temperate seas, in bays and estuaries. Some species live even in brackish or fresh waters. They are among the most numerous, if not the most numerous, of all marine fish.

Like the herring, anchovies live in large shoals. Within the smaller shoals the large individuals tend to be located below, the smaller individuals above, so allowing light to filter through. With larger shoals the formation is somewhat different. The anchovies separate out in sizes. It is believed that the larger individuals drive away the smaller, which shoal on their own. This characteristic shoaling behavior makes the anchovy especially valuable for canning, because the catch does not need to be sorted out for size.

Plankton feeders

Like the herring, the European anchovy is a plankton feeder. However, herrings select certain animals in the plankton, picking them up one at a time, whereas this anchovy swims forward with its mouth open, taking in small plankton more or less indiscriminately. The California anchovy feeds on the larvae of marine crustaceans such as euphausids, copepods and decapods. It uses random filter feeding but also pecks at prey.

The behavior of the shoals is determined by the feeding method. If a shoal of anchovies swam straight forward, those in front would capture all

The California or northern anchovy feeds both by filter feeding (straining water through its gills) and by pecking at its prey, the larvae of marine crustaceans.

Apart from commercial fishing, anchovies are mainly preyed upon by tuna. If a shoal of tuna approaches, the anchovies clump together. Only those on the fringes of the clump will be eaten.

the food. Instead, the leading individuals turn to either side and return to the rear of the shoal, so each gets its turn to feed. One result of this is that the shoal assumes the shape of a teardrop. When plankton is dense, however, the leading individuals fan out and the shoal assumes an oval shape.

Breeding

In the California anchovy spawning occurs from British Columbia to Magdalena Bay in Baja California. Spawning takes place throughout the year, both in- and offshore, but is mainly in winter and early spring. Spawning is mostly at depths less than 33 feet (10 m) and at temperatures of 50–55° F (10–13° C). Anchovies are oviparous, the eggs developing and hatching outside the mother's body. European anchovies spawn from April to November, with peaks usually in the warmest months. The eggs float in the upper 165 feet (50 m) and hatch after 24–65 hours.

Reaction to predators

The main predators of anchovies are tuna. The reaction of a shoal to their presence is to clump. The fish may be spread out, several hundreds across, but at the approach of the tuna the shoal contracts to form a sphere of thousands of fish, just a few feet across. Those with the least well-developed shoaling instinct remain on the fringes and are likely to be eaten by the tuna. Laboratory tests suggest that in any shoal there are a few individuals that panic less readily than the rest. In an alarm situation these stand their ground and the rest congregate around them. The anchovies detect tuna by the sound of the larger fish moving through the water.

CALIFORNIA ANCHOVY

CLASS **Osteichthyes**

ORDER **Clupeiformes**

FAMILY **Engraulidae**

GENUS AND SPECIES *Engraulis mordax*

ALTERNATIVE NAME
Northern anchovy

LENGTH
Average 10 in. (25 cm)

DISTINCTIVE FEATURES
Slender body, rounded in cross-section, with keel on belly; rounded, protuberant snout and very long jaw; translucent body; clear green on back, with bright silvery sides and silvery white belly

DIET
Larvae of marine crustaceans such as euphausids, copepods and decapods

BREEDING
Oviparous, development and hatching of eggs outside mother's body. Breeding season: all year, but mainly winter and early spring; number of eggs: not known; hatching period: 24–65 hours.

LIFE SPAN
Up to 7 years

HABITAT
Usually coastal waters within about 18½ mi. (30 km) from shore, but may be as far out as 300 mi. (480 km); also bays and inlets

DISTRIBUTION
Northeast Pacific: northern Vancouver Island, Canada, south to Cape San Lucas, Baja California, Mexico

STATUS
Abundant

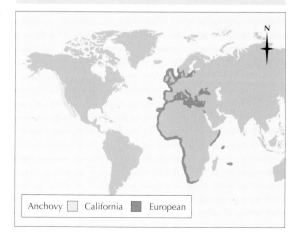

Anchovy ☐ California ■ European

ANEMONE

THE WORD ANEMONE, from the Greek for wind, was first used for a flower in 1551. At first these marine animals, which superficially resemble flowers due to their long tentacles and bright and varied colors, were called "plant-animals." The name sea anemone was not used until 1773. Today, marine zoologists almost invariably speak of them as anemones. That they are truly animals is not in doubt, although the superclass to which they were assigned is still called the Anthozoa, that is, plant-animals.

An anemone has simple sense organs, takes solid food and, surprisingly, is capable of locomotion. The most outstanding feature of anemones is the variety of their colors and, in many species, the striking and attractive patterns these make.

Many species

There are thought to be several thousand species of anemones, found worldwide. They live in the seas and oceans, from the tidal zone to depths of more than 33,000 feet (10,000 m). They are most abundant in warm seas where the largest and most colorful species are also found. The largest anemones are of the genus *Stoichactis* and can reach more than 3 feet (90 cm) across. In temperate seas, dense carpets of anemones may cover rocks exposed to strong currents, and one, *Metridium senile*, is often the most obvious colonizer of shipwrecks. The smallest sea anemones are little more than a pin's head in size.

Voracious feeders, anemones will eat any animal flesh they can catch and swallow, and they may swallow prey items larger than themselves. It is not unknown for one anemone to swallow another. They can, however, survive for a long time without food, gradually dwindling in size until quite minute. This may be one of the secrets of their longevity: anemones have been kept in aquaria for as long as 100 years.

Sedentary but mobile

Sea anemones seldom move but are by no means rooted to the spot. There are even burrowing species. Those that are normally seen fixed to a rock move by gliding on their base. Others somersault, bending over to take hold of the rock with their tentacles, then letting the base go and flipping this over to take hold beyond. A few species lie on their side to glide along, or inflate themselves, let go with their foot, and float away.

Simple anatomy

An anemone is a cylindrical bag with a ring of tentacles surrounding the oral disc, or mouth, on the upper surface. The opposite end is flattened and forms a basal disc, or foot, by which the animal sticks to a solid support. Its interior is one large stomach, sub-divided by curtains of tissue, or mesenteries, which hang down, partially dividing the stomach into eight compartments.

The body wall of an anemone is made up of two layers of cells. There is, however, a set of longitudinal muscles running from the foot to the tentacle bases, and a set of circular muscles running around the body. By the lengthening and contraction of these muscles the body can be drawn out or pulled in. There is also a series of retractor muscles that assist in the sudden withdrawal of body and tentacles.

Stinging tentacle feeders

Food is caught by the tentacles, which are armed with stinging cells. When a small animal, such as a shrimp or a fish, touches a tentacle the stinging cells come into action, paralyzing and holding it. Adjacent tentacles bend over and continue to hold and sting the prey, eventually drawing it into the mouth.

The stinging cells, or nematocysts, are double-walled capsules, filled with poison, set in the outer surface of the tentacles. Each contains a coiled hollow thread, sometimes barbed at the base. At the outer end of the capsule is a thornlike trigger. When this is touched, the coiled thread is

Anemones feed on almost any animal matter they can catch with their tentacles. Pictured is a gem anemone, Bunodactis verrucosa.

ANEMONES

PHYLUM	**Cnidaria**
SUPERCLASS	**Anthozoa**
CLASS	**Hexacorallia**
ORDER	**Several, but mainly Actiniaria**
GENUS	**Actinia, Anemonia, Stoichactis, Metridium and many others**
SPECIES	**Several thousand species**

ALTERNATIVE NAME
Sea anemone

LENGTH
Typically ¾–4 in. (2–10 cm) across; larger species up to 3¼ ft. (1 m) across

DISTINCTIVE FEATURES
Flowerlike appearance due to many feeding tentacles, usually in concentric rings around mouth; often brightly colored

DIET
Small animals and pieces of detritus

BREEDING
Great variety of sexual and asexual breeding methods, including budding and splitting

LIFE SPAN
Many species live for a long time, perhaps up to 100 years or more

HABITAT
On rocks, stones and seaweed or buried in sand and gravel; from intertidal zone out to deep sea

DISTRIBUTION
Worldwide, though not common in estuaries or intertidal, sandy areas

STATUS
Many species common

Beadlet anemones exposed at low tide, with their tentacles retracted. These anemones can be extremely aggressive, often inflicting serious damage on one another.

shot out. It turns inside-out as it is ejected, its fine point pierces the skin of the prey and the paralyzing poison flows down the hollow thread. Some kinds of nematocysts stick to the prey instead of piercing the skin, and in a third type the thread wraps itself around the victim. Some nematocysts are triggered by the presence of certain chemicals as well as by touch.

Some aggressive species

Predators of anemones include large sea slugs, sea spiders, fish and sometimes starfish and crabs. The common beadlet anemone, *Actinia equina*, can be very aggressive to members of its own species and fights lasting several minutes have been observed. The loser can often be quite visibly damaged by the stinging cells of the victor, and usually moves away. It has been found that the more common red variety almost always triumphs over the less common green variety.

Sexual and asexual reproduction

Anemones display a great variety of sexual and asexual breeding methods. Most anemones are either male or female, but some are hermaphroditic. In some, eggs and sperm are shed into the surrounding water, in others the larvae develop inside the parent body. The fertilized eggs sink to the bottom and divide, or segment, to form oval larvae. These move about the seabed but finally each comes to rest, fastens itself to the bottom, grows tentacles and begins to feed.

Other species, for example in the genus *Anemonia*, split longitudinally to form separate individuals, or grow a ring of tentacles halfway down the body, after which the top half breaks away to give two anemones where there was one

before. Alternatively, young anemones may be formed by fragmentation, or laceration. In fragmentation small anemones, complete with tentacles, arise from the base of a parent, become separated and move away. Laceration occurs in some of the more mobile species, such as those of the genus *Metridium*. As the anemone glides over the rocks, pieces of the base are ripped away and, being left behind, regenerate to form minute but otherwise perfect anemones. In other species young anemones are formed asexually inside the parent and are subsequently spat out as perfectly formed minatures.

ANGELFISH

THE NAME ANGELFISH is used for three types of fish, which often causes confusion. The first is a relative of sharks, which we treat under its usual name, monkfish. The other two are bony fish, one of which is marine, the other freshwater. The freshwater type is a member of the cichlid family, discussed elsewhere.

This article refers to all the fish that were, until the mid-1970s, grouped together in the family Chaetodontidae. They include both marine angelfish and butterflyfish, together numbering more than 150 species. The two share many characteristics, but an angelfish has a spine on each gill cover, which a butterflyfish lacks. This difference has caused scientists to distinguish between the two by referring to angelfish under the separate family name Pomacanthidae, which includes 9 genera and about 74 species. To confuse matters even further, the term butterfly fish is also used to refer to a single species of freshwater fish, *Pantadon buchholzi*, discussed elsewhere.

Most angelfish are small, up to 8 inches (20 cm) long, but some, like the French old woman angelfish, *Pomacanthus striatus*, can reach a length of 1¾ feet (50 cm). The outline of the body, because of the well-developed fins, is shaped like an arrowhead.

Colorful and curious

Angelfish and butterflyfish live mainly in pairs or small groups around reefs, rocks or corals in shallow seas. A few enter estuaries. They do not dash away as most fish do when, for example, a diver intrudes into their living space. They move away slowly, occasionally tilting the body to take a closer look at the newcomer.

The most outstanding feature of these fish is the wide range and beauty of their colors and patterns. In many species the young fish have the same colors as the adults, but in others the differences are so great that it looks as if they are two different species. Their behavior is also different. When quite small (up to a few inches long) they tend to be solitary. Individuals are usually found in the same places day after day, usually near some kind of shelter into which the fish can dart when it senses danger. The shelter may be under a rock bed or among seaweed, or even a tin can lying on the seabed. In an aquarium the younger

fish will be aggressive toward each other, but one kept on its own readily becomes tame and learns to feed from the hand.

One of the most striking angelfish is the rock beauty, *Holacanthus tricolor*. It is jet black in front and yellow in its rear half, and its fins are bright yellow with red spots. It appears to be strongly curious, drawing near to any swimmers. The French angelfish, *Pomacanthus semicirculatus*, is another striking fish. It is black with contrasting yellow vertical bands and a yellow face.

Feeding

Angelfish have small mouths with many teeth to crush the small invertebrates on which they feed. In some species the snout is somewhat elongated so that it can be inserted into cracks in rocks or coral to capture animals for food.

In certain butterflyfish species, such as the copperband butterflyfish, *Chelmon rostratus*, the snout is long and tubelike with the small mouth at the end. This enables the fish to probe even deeper into the coral rock crevices for their food.

Parental care

Little is known of the breeding habits of angelfish, but they are probably similar to those of the butterflyfish, which take great care of the eggs and fry (young fish). Both the male and

This brilliantly colored butterflyfish of the genus Hemitaurichthys *displays the enlarged flaplike or winglike fins that have suggested the names of both the angelfish and the butterflyfish.*

The emperor angelfish, Pomacanthus imperator, lives in the tropical waters of the Indian and Pacific Oceans.

female clean a patch of flat rock, where the female lays her eggs. The male swims close over them, shedding sperm for fertilization. The eggs are tended for 4 to 8 days by the parents before the fry hatch and sink to the bottom.

The parents guard the fry until they are free-swimming enough to hide in crevices and weed. The fry are unlike the adults in that their bodies are long and slim. They do not assume full adult shape before 3 or 4 months have passed.

The purpose of color

We know that animals use colors and color patterns as camouflage to hide from predators or to steal close to prey undetected. We also know that conspicuous colors, especially combinations of yellow, black and red, are used to warn other animals that their possesser is poisonous, bad-tasting or has a sting. The colors of angelfish certainly fail to hide their wearer, and there is no indication that angelfish are poisonous, inedible, or have a sting. However, it is thought that their conspicuous colors might be used as a sign of aggression and to mark their territory.

Experimentally, a mirror was placed in an aquarium with a French angelfish. The fish drew near, then threw itself sideways and flicked its bright blue pectoral fins like flashing signals. This suggests that they use their colors both to advertise possession of a territory as well as to warn off an intruder of their own species.

QUEEN ANGELFISH

CLASS	**Osteichthyes**
ORDER	**Perciformes**
FAMILY	**Pomacanthidae**
GENUS AND SPECIES	***Holacanthus ciliaris***

WEIGHT
Up to 3½ lb. (1.6 kg)

LENGTH
Up to 17½ in. (45 cm)

DISTINCTIVE FEATURES
Adult: 1 spine on each gill cover; body is shaped like an arrowhead; brilliant yellow tail and pectoral fins; bright blue spot on nape; patches of violet and red on various parts of body. Juvenile, when under 2 in. (5 cm) long: dark brown to black overall; 3 light blue vertical bars on sides and 1 on each dorsal fin, which disappear with age.

DIET
Adult: mainly sponges, supplemented by small amounts of algae, tunicates and bryozoans. Young: also pick ectoparasites from other fish.

BREEDING
Hermaphroditic (having both male and female reproductive organs) with a social system similar to harems

LIFE SPAN
Not known

HABITAT
Coral rocks and reefs among sea fans, sea whips and corals

DISTRIBUTION
Tropical western Atlantic from Florida and Gulf of Mexico south to Brazil; also in east-central Atlantic at St. Paul's Rocks

STATUS
Common

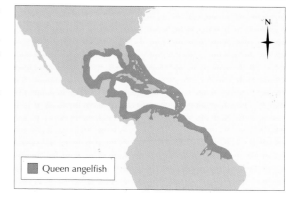

Queen angelfish

ANGLERFISH

THERE ARE SOME 297 species of anglerfish belonging to 16 families and about 65 different genera. These fish are named for their characteristic habit of keeping still and fishing for their prey, usually small fish. The rod of the anglerfish is a modified spiny ray of the dorsal fin with a fleshy bait at the tip. Habitual immobility means little expenditure of energy, and less need for breathing. This is reflected in the small gills of anglerfish, with only a small gill opening. *Pediculati*, the old name for anglerfish, means "small foot." This refers to the elbowed pectoral fins the fish use rather like feet to move over the seabed in short jumps. The pelvic fins are also somewhat footlike, but they are small and hidden on the undersurface in front of the pectoral fins.

Anglerfish can be divided into the following groups: goose-fish (family Lophiidae), frogfish (Antennariidae), handfish or warty anglers (Brachiomichthyidae), batfish (Ogcocephalidae) and deep-sea anglerfish (the remaining families, discussed elsewhere). A well-known species is the angler or monkfish, *Lophius piscatorius*. It is up to 6½ feet (2 m) long, with a large, flattened head, and has a wide mouth. It should not be confused with the 13 sharks known as monkfish, described under that name.

Lying in wait

Anglerfish are found at all depths throughout tropical and temperate seas. Most species live near the seabed, their bodies ornamented with a variety of warts and irregularities, as well as small flaps of skin. These, along with their usually somber colors, serve to camouflage the fish as they lie immobile among rocks and seaweed. The sargassum angler, *Histrio histrio*, has a specialized camouflage. It lives exclusively among floating sargassum the weed of the world's oceans, and uses its pectoral fins to grasp the weed, keeping its stationary position. Deep-sea anglerfish differ in that they are not bottom-living, but instead swim feebly about.

Angling for food

Anglerfish feed by attracting small fish using some form of lure. In the goosefish and monkfish this is a "fishing rod" bearing a fleshy flap at its tip, which is waved slowly back and forth near the mouth. In other species the rod lies hidden, folded back in a groove, or lying in a tube, and is periodically raised or pushed out and waved two or three times before being withdrawn.

The lure at the end of the rod is often red and wormlike in shape. The anglerfish does not allow its quarry to take the bait. Instead, the lure is waved until a fish draws near, then it is lowered toward the mouth. As the victim closes in on it, the rod and its lure is suddenly whipped away. The fish opens its huge mouth, water rushes into this capacious landing net and the prey is sucked

Anglerfish are named for their habit of fishing for prey. They attract passing small fish using a modified spiny ray tipped with a fleshy lure or bait.

MONKFISH

CLASS	**Osteichthyes**
ORDER	**Lophiiformes**
FAMILY	**Lophiidae**
GENUS AND SPECIES	***Lophius piscatorius***

ALTERNATIVE NAME
Common goosefish

WEIGHT
Up to 66 lb. (30 kg)

LENGTH
Up to 6½ ft. (2 m)

DISTINCTIVE FEATURES
Fishing rod (modified ray of dorsal fin) with fleshy bait at tip; head and front part of body large, wide and flattened from above; large, semicircular mouth; lower jaw longer than upper jaw; fringed lobes around sides of head and body rather than scales; loose skin; brownish, reddish or greenish brown in color with darker blotches; white underside

DIET
Fish such as flatfish, haddock and dogfish; occasionally small seabirds

BREEDING
Breeding season: late winter, spring and summer; number of eggs: up 1.5 million

LIFE SPAN
Not known

HABITAT
Sandy and muddy seabeds at depths of 65–3,280 ft. (20–1,000 m); also on rocky grounds

DISTRIBUTION
Eastern and North Atlantic; Mediterranean; Black Sea

STATUS
Locally common

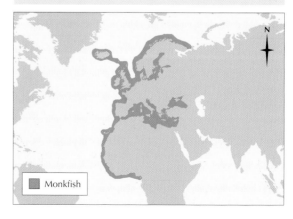

Monkfish

One of the best known species of anglerfish is Lophius piscatorius, *also known as the monkfish. It lies half-buried in the sediment waiting for its prey.*

in, after which the mouth snaps shut. The monkfish uses this method to prey not only upon flatfish, haddock and dogfish, but also occasionally upon small seabirds.

Batfish have a somewhat different fishing method. Often they are covered with outgrowths of skin that look like small seaweeds and polyps known as sea firs. Small fish swim near and try to nibble the flaps of skin. The batfish will then produce its rod and bait to lure the fish closer until they are caught.

Breeding

Some deep-sea species of anglerfish show a peculiar relationship between male and female. The dwarf male, about ⅓–½ inch (1–1¼ cm) long, attaches itself to the female, which may be up to 3¾ feet (1.1 m) in length. The male becomes so securely attached that the two grow together, even sharing a blood system. The female is then, in effect, a self-fertilizing hermaphrodite, the male being reduced to a mere sperm-producing organ.

Another outstanding feature of the breeding cycle of some anglerfish is the size of the egg masses. Female goosefish and frogfish lay eggs in a jellylike mass, up to 40 feet (12 m) long and 3 feet (90 cm) in width. This mass floats at the surface. The relatively large, pear-shaped eggs are attached by the narrow end to this mass, which floats at the surface. There may be nearly 1.5 million eggs.

The larva begins to develop black pigment even before it leaves the egg. The larva is in an advanced stage when hatched and already has the beginnings of its fishing rod. Later, other spines develop on the back and branched fins grow down from the throat, so the larva looks very unusual. After a free-living existence, the young take up bottom-living when about 2–2⅓ inches (5–6 cm) in length.

ANOLE LIZARD

ANOLE LIZARDS ARE ANY of approximately 250 species in the genus *Anolis*, one quarter of the total number of species in the iguana family of lizards. Because the genus is so large, some authorities split it. Under this scheme, the green anole of the southeastern United States remains in the genus *Anolis*, but the Jamaican anole is placed in another genus, *Norops*.

The anoles' heads are triangular with elongated jaws, and their bodies are slender, ending in a long, whiplike tail. Like the geckos, the toes have sharp claws as well as adhesive pads. These enable the anoles to climb tree trunks and sheer walls. Males have a flat throat sac, or dewlap, which they expand by muscular action when they are excited. This expands the folds of skin to reveal a pattern of colors between the scales, often green, red, white, yellow and black in many combinations.

Anoles are small lizards, most ranging in length from 5 to 10 inches (13–25 cm). A few species are larger than this, the knight anole, *Anolis equestris*, of Cuba being 18 inches (45 cm) in length. Of this, two-thirds is tail, so the knight anole is by no means a large lizard. It is, however, a striking reptile: pale green with white markings on its body, a braided yellow pattern on its head and patches of blue around the eyes. The male's throat sac is pale pink.

The best-known anole is the green anole, *A. carolinensis*, from the southeast of the United States. It is about 6–7 inches (15–18 cm) long, with a pale green body, the male having a throat sac spotted with red and white. The leaf-nosed anole of Brazil gets its name from the sideways, flattened structure that projects beyond its snout for a distance equal to the length of its head.

Tree-dwelling lizards
Anole lizards are found only in the Americas, where they range from North Carolina to southern Brazil and Chile, and are particularly abundant in the Caribbean. Most of them live in trees, running along the branches with the aid of their long, delicate toes and adhesive pads. A few species have enlarged toe pads that act as tiny parachutes, enabling them to jump from considerable heights. Other anoles have become associated with humans, living in houses and gardens where they often become quite tame.

Three of the more unusual anoles (*A. lionotus*, *A. poecilopus* and *A. barkeri*) live along the banks of streams, diving into the water and hiding under stones when frightened. Living in Cuba are two cave-dwelling anoles. One of these, a pale, translucent lizard with brick-red stripes running across its body, lives in limestone caves frequented by bats. Anole lizards are mainly insect eaters, but will take fruit and plant material when it is abundant.

Able to change color
The green anole is commonly called the American chameleon because, like most species of anoles, it is adept at changing color. Experiments and observations in natural conditions show that anole lizards change color mainly in response to temperature and light intensity. Background color (the principal factor in chameleon color changes) will affect the color of an anole to some extent, but if it is kept cool, at about 50° F (10° C), it will go brown, whatever the background. If the temperature is raised to 70° F (21° C) it turns green, but only so long as the light is dim. If the light is bright it stays brown. In normal conditions the green anole tends to be green at night and brown by day. Color changes in anoles are also sometimes associated with aggression. For example, the male green anole becomes bright green if he wins a fight but brown if he loses.

Anole lizards change color in response to changes in temperature or light intensity. This compares with the chameleons, in which background color is the most important factor.

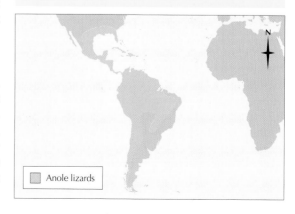

ANOLE LIZARDS

CLASS	**Reptilia**
ORDER	**Squamata**
SUBORDER	**Sauria**
FAMILY	**Iguanidae**
GENUS	***Anolis***
SPECIES	**About 250 species**

ALTERNATIVE NAME
Green anole: American chameleon

LENGTH
Most species: 5–10 in. (13–25 cm); a few species larger than this

DISTINCTIVE FEATURES
Small size; triangular head; elongated jaws; long claws; most species have adhesive pads on their digits; extensible dewlap (throat sac) on underside of throat (male only); most species can change color

DIET
Mainly insects; also plant material

BREEDING
Varies according to species. Hatching period: 42–70 days.

LIFE SPAN
Most species: 2–3 years in the wild, or up to 6 years in captivity

HABITAT
Most species in trees; a few species in caves and suburban areas

DISTRIBUTION
Americas from North Carolina south to southern Brazil and Chile; also on many Caribbean islands

STATUS
Generally abundant; on some Caribbean islands anoles achieve densities greater than for any other lizards worldwide

Anole lizards

Male anoles, such as this brown anole, Anolis sagrei, *have a brightly colored throat sac that they expand during territorial displays or courtship.*

The mechanism of color change varies among reptiles. For example, in chameleons, the pigment-containing cells in the skin, responsible for color changes, are controlled by nerves. As a result, chameleons can change color quite rapidly. However, in anoles the cells are controlled by a hormone, a chemical messenger called intermedin, which is secreted into the blood by the pituitary gland. This is a slow process compared with the action of the nervous system, and anoles can take up to 10 minutes to change color.

Breeding behavior

The male anole is larger and more brightly colored than the female. He holds a territory that he defends against other males by displaying the colors of his throat sac and, at times, by fighting. The throat sac is also used to attract the female which, if willing, turns her head to one side. The male approaches her from the rear, grabs her neck with his jaws, slips his tail under hers, and mates. The eggs are nearly always laid in the ground, the female coming down the tree to dig a hole with her snout. She lays the eggs into the hole, which she then fills in. The cave anoles lay their eggs in narrow crevices in the cave walls, or between stalactites. Anoles often lay only one or two eggs at a time. The eggs, which are not guarded, hatch after 42–70 days.

Many enemies

Predators of anoles include hawks, cats and mongooses. In one experiment, 200 anoles were marked and released. A year later only four had survived. Like many abundant animals, there is a very rapid turnover of population, very few anoles even reaching maturity. The maximum age an anole can reach in captivity is over 6 years, but they probably live only 2 or 3 years in the wild.

ANTBIRD

THERE ARE ABOUT 260 species of antbirds, arranged in 53 genera. Not many have been studied in any great detail, and some are known only from skins brought back by collectors. The name antbird was formerly used for what was considered to be a single, very diverse family. This has since been divided into two families: the Thamnophilidae, the typical antbirds, and the Formicariidae, the ground antbirds. Thamnophilidae contains about 75 percent of all species. Many members of these families have been given common names with the prefix "ant" followed by the names of unrelated birds that they resemble in a particular characteristic. Thus there are ant-thrushes, antshrikes, antcreepers, antwrens, antvireos and several others. Other antbirds have been given names relating to their appearance or habits, such as bare-eye, fire-eye and bushbird.

Antbirds range in size from the 3–3¼-inch (7.5–8.5-cm) pygmy antwren, *Myrmotherula brachyura*, to the 12-inch (30-cm) antpittas. Their bills are usually hooked, particularly strongly in the antshrikes. The antbirds' plumage is generally dull but is often enlivened with patches of black and white. In many species there is a marked difference between the sexes, with the black parts of the male's plumage being replaced by brown or chestnut in the female. The male of the white-flanked antwren, *Myrmotherula axillaris*, is almost wholly black, with only a few spots of white on its wings and, on the flanks, white patches that are exposed during display. In contrast, the female is olive gray. This sexual dimorphism, as it is called, is not found in the ant-thrushes and antpittas.

Prefer dense cover

Antbirds are found on the mainland of Central and South America, being absent only from Chile and much of Argentina. By far the greatest variety occurs in the Amazon Basin, with some 30 or 40 species sympatric (occupying the same range) at some localities. They are also found on Trinidad and Tobago. Antbirds live in thickets and forests on lowlands and on mountain ranges, but never in open country or plains.

Antbirds are extremely vocal, and some of them are best recognized by their song, which is often very shrill. Ant-thrushes have a mellow, resonant whistle that has been described as being almost wistful or melancholy.

As would be expected of birds living in dense cover, antbirds are not strong fliers. Their wings are short and rounded, and they do not migrate. With so many species occupying a relatively small region, every part of it is exploited. The black-faced ant-thrush, *Formicarius analis*, is ground-living and escapes danger by running, taking off only as a last resort. The antpittas also live on the forest floor. These birds have long legs and they stalk around twitching their short, frequently cocked tails.

The spotted antbird, *Hylophylax naevioides*, lives just above the forest floor and is rarely seen more than 6 feet (1.8 m) from the ground. Antvireos and antwrens tend to be more arboreal (tree-living). The white-flanked antwren is found in the middle layers of the forest, never on the ground or up in the topmost branches. The barred antshrike, *Thamnophilus doliatus*, is the best known of the antbirds because, while most of the others avoid crossing open ground, this antshrike is often found in clearings, forest margins, forest regrowth and bushes or gardens around human settlements. Another antshrike, the black-crested antshrike, *Sakesphorus canadensis*, occurs in arid scrub, savanna and gallery forest.

Solitary birds

Most antbirds are solitary or go about in pairs, but a few are nearly always seen in tight groups and the white-flanked antwren goes around in mixed flocks of other, unrelated species such as vireos, cotingas and woodcreepers, as well as with other antbirds.

Amazonian antshrike, Thamnophilus amazonicus, in the Peruvian rain forest. Antshrikes resemble true shrikes in some physical characteristics and feeding habits.

The name antbird is derived from some species' habit of following army ant columns to feed off the insects that flee before these voracious ants. Pictured is a female great antshrike, **Teraba major.**

Following ant columns

The name antbird is actually quite misleading because these birds do not habitually feed on ants. The ant-thrushes and antpittas, rather, flick through the leaf litter with their bills to look for snails and beetles. Some of the larger species occasionally eat small reptiles and may take the eggs and young of other birds. The white-flanked antwren, meanwhile, forages for insects hiding in the bark of twigs. Nonetheless, about 15 species, in 6 genera, do live up to their name in that they are so-called professional followers of army ant swarms. They feed mainly on arthropods and insects fleeing the ant columns, which form from mid-morning to mid-afternoon. Other birds also do this, particularly woodcreepers.

Breeding poorly known

Breeding behavior is very poorly understood for most antbirds, although a little is known about certain species. For example, the spotted antbird is thought to breed in the first two months of the rainy season. The nest is built near the ground by both adults, each bird arranging its own material in the nest. In most antbirds, the nest is cup-shaped. Some of the ant-thrushes are unique among antbirds in making nests in holes in trees. It is common among antbirds for the male to feed his mate during courtship or while she is on the nest. Two eggs are normally laid and are brooded by both parents during the day, the female only at night. The eggs hatch after 14–20 days and the chicks are brooded and fed at the nest for 9–18 days.

Variations on a theme

In some cases the names of the antbirds reflect habits and characteristics of their namesakes. The antvireos, for example, have short bills and search leaves and twigs for insects as do true

BLACK-FACED ANT-THRUSH

CLASS	**Aves**
ORDER	**Passeriformes**
FAMILY	**Formicariidae**
GENUS AND SPECIES	*Formicarius analis*

LENGTH
Head to tail: 6¾–7 in. (17–18 cm)

DISTINCTIVE FEATURES
Compact body; large head; short, frequently cocked tail; long legs; gray underparts; brown upperparts; black throat

DIET
Arthropods and insects; occasionally small snakes, lizards and frogs

BREEDING
Very poorly known. Number of eggs: probably 2; incubation period: 20 days.

LIFE SPAN
Not known

HABITAT
Forest floor in tropical rain forest and in humid, secondary woodland

DISTRIBUTION
Mexico south to Amazon Basin, including islands of Trinidad and Tobago

STATUS
Fairly common

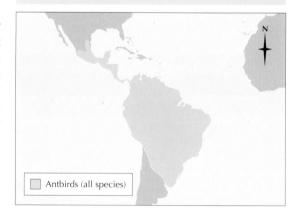

Antbirds (all species)

vireos. They also build vireolike nests in the forks of branches. The antshrikes have well-developed hooks on their bills and take large insects and small reptiles in the same way as shrikes, which the males also resemble in their black-and-white plumage. Elsewhere, the relationships between antbirds and their namesakes are more obscure. Ant-thrushes resemble thrushes only in plumage and antwrens are merely the smallest of antbirds.

ANTEATER

With its long, tubular snout, long, bushy tail and inturned front feet, the giant anteater has a most unusual appearance. The largest of four species of anteaters from three different genera, the giant anteater is about 3½–4 feet (1–1.2 m) in length, with a 2½–3 foot (75–90-cm) tail. It might weigh up to 85 pounds (38.5 kg). Sometimes called the ant bear, its hair is coarse and stiff, gray brown on the head and body, becoming darker on the hindquarters and tail. Across each shoulder is a wedge-shaped, black stripe bordered with white, giving the animal an effective camouflage. Its mouth opening is small and its tongue is extremely long.

The three other species of anteaters, the silky or two-toed anteater (*Cyclopes didactylus*) and the northern and southern tamanduas (*Tamandua mexicana* and *T. tetradactyla*), are not well known. They are a little more conventional in appearance than the giant anteater, with shorter snouts and no plume of long hair on the tail, but they also have long tongues for feeding. The silky anteater is so called because of its soft, silky, yellowish coat. It is the smallest species, its head and body

being just 6–9 inches (15–23 cm) long. Its prehensile tail is a little longer than this. The tamanduas, also called collared or lesser anteaters, also have a prehensile tail, which is naked for most of its length. Their fur is usually tan with a black waistcoat around the shoulders and body. Some individuals are entirely tan or entirely black.

Anteaters are grouped in the order Edentata, which means "without teeth." In rare cases anteaters do have a few teeth, but they and the other members of this order, the sloths and armadillos, are mostly toothless.

Shuffling walk

Anteaters live in Central and South America. The giant anteater is found in swamps, open forests and savannas, from southern Panama to southern Brazil. It is thought that it is now extinct in Belize, Guatemala and El Salvador, and it is considered vulnerable. Both the silky anteater and the tamanduas live in forests, from southern Mexico south to Bolivia and Brazil. They are rarely seen and there are few estimates of their populations remaining in the wild.

Dry grasslands such as the Pantanal of Brazil (below) are ideal for the giant anteater. With its long snout, bushy tail and shuffling gait, the giant anteater is unmistakable.

The giant anteater lives on the ground, shuffling about with its nose down in an almost continual search for food. When alarmed it is able to run quite quickly. It often has no permanent resting place or burrow but curls up, with its head between its forelegs and tail wrapped round its body, in any sheltered place. It sometimes takes over the abandoned burrow of another animal. In remote areas, giant anteaters are diurnal (day-active), but they have become largely nocturnal near towns.

Unusual anatomy

In addition to its nose, tongue and tail, the giant anteater also has most unusual feet. It walks on the knuckles of its front feet, giving the animal an almost crippled appearance. This stance appears to be an adaptation to protect the long, sickle-shaped claws from being blunted. The awkwardness of the front feet is emphasized by the nearly human shape of the hind feet. The heel rests on the ground and the five toes are of almost equal length.

Little is known about the habits of the northern and southern tamanduas and the silky anteater. They are arboreal rather than ground living, their prehensile tails being used in climbing. They are also mainly nocturnal, spending the day curled up in a hollow tree or on a branch. The silky anteater's feet are similar to those of the giant anteater, but there is an extra joint in the hind feet that enables it to bend its toes back under the sole to give a good grip on branches.

Silky anteaters are acrobatic nocturnal residents of Central and South American rain forests. They are named for their unusually soft coats.

GIANT ANTEATER

CLASS **Mammalia**

ORDER **Edentata (alternatively Xenarthra)**

FAMILY **Myrmecophagidae**

GENUS AND SPECIES *Myrmecophaga eridactyla*

ALTERNATIVE NAME
Ant bear

WEIGHT
45–85 lb. (20–38.5 kg)

LENGTH
Head and body: 3½–4 ft. (1–1.2 m); tail: 2½–3 ft. (75–90 cm)

DISTINCTIVE FEATURES
Large size; extremely long, tubular snout with equally long tongue; powerful forelimbs with inturned feet and strong, sickle-shaped claws; coarse, gray brown coat with wedge-shaped, black stripe across each shoulder; long, shaggy tail

DIET
Almost exclusively ants and termites

BREEDING
Poorly known. Breeding season: usually March–May; gestation period: about 190 days; number of young: 1; breeding interval: not known.

LIFE SPAN
Up to 26 years in captivity

HABITAT
Savanna, grassland, swamps and dry, open forests

DISTRIBUTION
Panama south to southern Brazil and eastern Paraguay; probably extinct in rest of Central America

STATUS
Vulnerable; populations declining in many parts of range

Giant anteater

A northern tamandua in rain forest, Costa Rica. Unlike the other anteaters, tamanduas eat bees and honey as well as termites.

Sticky tongues

Giant anteaters feed almost exclusively on ants and termites and their soft-bodied grubs. They mostly avoid aggressive insects. To get to the ants they tear holes in the tough walls of ant nests with the sharp claws of their front feet. The long muzzle is then pushed in and the 8–10-inch (20–25-cm) tongue probes around the galleries of the nest to trap the insects with its sticky saliva. Giant anteaters are not as destructive as aardvarks. Often they open a nest but feed for only a minute or two before moving on; this behavior helps to preserve a steady food supply.

The silky anteater has a similar diet, feeding mainly on ants. Tamanduas often eat termites and also feed on bees and honey.

Secretive breeding

Anteaters are solitary animals, and when two are seen together it is most likely that they are a mother and her single offspring. Giant anteaters breed in the fall, from March to May. After a gestation period of 190 days, the young is born in the spring. The female gives birth standing up, using her tail as a tripod for support.

All anteater species produce a single young. In the case of the giant anteater, the baby clings to its mother's back until it has grown to almost half her size, staying with her until she is pregnant again. It suckles for around 6 months. Tamanduas are also thought to mate in the fall, with the young born in spring. However, the gestation period is shorter in these species, at between 130 and 150 days. Little is known of breeding in the silky anteater. The mother is thought to make a nest of dry leaves in a hollow tree and to leave her baby there while she goes out feeding. She is also said to feed her baby on a regurgitated mush of insects. Feeding young by regurgitation is common among birds but rare in mammals, in which young are normally suckled on milk until they can take solid food.

Formidable adversaries

The main predators of the silky anteater are the harpy eagle (*Harpia harpyja*), the spectacled owl (*Pulsatrix perspicillata*) and the various species of hawk-eagles. Its first line of defense is camouflage. When it is around silk-cotton trees, the sheen of the silky anteater's fur makes it very hard to see among the mass of silverish fibers that cover the seed pods. If it is attacked, the silky anteater rears up on its hind legs, slashing at its adversary with the long claws of its front feet. The giant anteater is also a formidable animal, deterring jaguars and other carnivores with its claws. Tamanduas have a similar method of defense.

ANTELOPES

A REMARKABLY DIVERSE AND successful group, the antelopes are thought to have originated from a small, gazelle-sized ancestor, *Eotragus*, which lived at the forest margins in Europe and North Africa about 20 million years ago. Antelopes are found in the equivalent niches to deer (Cervidae), but occur only in Africa, the Middle East and Asia. They range in size from the tiny royal antelope, *Neotragus pygmaeus*, weighing only 3 pounds (1.5 kg) and standing just 8 inches (20 cm) at the shoulder, to the giant eland, *Taurotragus derbianus*, which weighs up to 2,200 pounds (1,000 kg) and stands up to 6 feet (1.8 m) high.

The classification of antelopes is still much debated by zoologists, but the general consensus is that there are about 100 species divided between nine different tribes. Three-quarters of these are from Africa and the remainder from Asia. Strictly speaking, the name antelope applies to species within one tribe, the Antilopini, after the Indian blackbuck, *Antilope cervicapra*, one of its members. But more generally, the antelopes include all members of the family Bovidae, except the sheep and goats (the tribe Caprini), buffalo and cattle (Bovini) and goat-antelopes (Rupicaprini).

The importance of horns

The most striking feature uniting the antelopes is the presence of horns in the males. Except in the four-horned antelope, *Tetracerus quadricornis*, and the nilgai, *Boselaphus tragocamelus*, both with two pairs of horns, all males have two horns. These range in size and structure from short spikes of about ¾ inch (2 cm) in length to large, elaborate spirals. Horns play an important role in fights between males, which take place to establish dominance in herds or territory ownership. In addition, their shape affects the form of combat and fighting tactics. Females also possess horns in nearly half the antelope species, although they are usually thinner and shorter than those of the males. In both sexes, the horns begin to appear during the first year, and develop gradually until after the animal reaches sexual maturity.

Antelope horns consist of outer sheaths of keratin, the horny substance from which human fingernails are made, that grow over bony extensions from the frontal bones of the skull. The horns are permanent and do not regenerate, and in older animals may be worn or broken. These traits help to distinguish antelopes from related groups such as deer and from the pronghorn antelope, *Antilocapra americana*, of North America, which is not in fact a true antelope. Deer antlers are composed of solid bone and are shed annually, while the pronghorn antelope sheds its keratinous sheath but retains the bony core when the mating season finishes. The pronghorn is also distinctive for being the only "antelope" in North America.

A female kob with her young. The kob is a grazing antelope, feeding mainly on grasses and grasslike plants.

CLASSIFICATION
CLASS Mammalia
ORDER Artiodactyla
FAMILY Bovidae
SUBFAMILY Bovinae; Hippotraginae; Cephalophinae; Antilopinae
TRIBE Strepsicerotini; Boselaphini; Reduncini; Alcelaphini; Hippotragini; Cephalophini; Antilopini; Neotragini; Saigini
NUMBER OF SPECIES Approximately 100

Like deer and the pronghorn antelope, all antelopes have cloven hooves, formed by the third and fourth digits. Depending on the species, the first and second digits have either been lost during evolution or are retained as small dew hooves (vestigial hooves, not reaching to the ground).

Feeding habits and social organization

Antelopes are ruminants: their stomach contains four chambers, one of which is known as the rumen. In common with other ruminants, such as buffalo, cattle, sheep and goats, they swallow food after chewing it briefly. The larger food fragments are stored in the rumen and regurgitated later during resting time for further chewing. This form of digestion increases the nutritional value of plant foods.

The social organization of female antelopes is largely dependent on their feeding habits. Smaller antelopes, such as the klipspringer (*Oreotragus oreotragus*), suni (*Neotragus moschatus*), dik-diks (genus *Madoqua*), and duikers (genus *Cephalophus*), feed largely on trees. For this reason they are referred to as browsers. They prefer feeding on the young leaves, flowers and fruits of specific tree species. This food is localized and can be seasonally scarce, so females live alone or with their offspring and sometimes defend territories at good feeding sites. Meanwhile the males defend territories that overlap with the range of one or more females. Small antelope species such as these are either nocturnal or depend on cover to avoid detection by predators.

Larger antelope species tend to feed on a broad range of plants. They may be both browsers and grazers, such as the reedbucks (genus *Redunca*), and the impala (*Aepyceros*

Young males, such as these Thomson's gazelles, often live in bachelor herds so as to provide protection from predators.

melampus), or pure grazers, such as the plains-living gazelles (genus *Gazella*). As grasses and grasslike plants are more widely distributed than trees and shrubs, the defense of feeding territories is not worthwhile and group-living is possible for the grazing gazelles. The groups, or herds, number three to five individuals in the case of reedbucks and the beira antelope, *Dorcotragus megalotis*, or up to several thousand in the saiga antelope, *Saiga tatarica*, and wildebeests, genus *Connochaetes*. Group-living provides additional protection from predators because responsibility for vigilance can be shared between members of the herd.

Antelopes rarely defend themselves against predators, although some of the larger species such as the sable antelope, *Hippotragus niger*, have on occasion been seen to stand their ground. The usual response to attack is rapid flight. Thomson's gazelle, *Gazella thomsoni*, initially attempts to avoid being chased by stotting, a stiff-legged, jumping display designed to advertise the individual's good condition and fitness. Meanwhile, the mountain-inhabiting klipspringer uses steep, vertical slopes to escape.

Breeding

Female antelopes have one or sometimes two pairs of mammae (teats) and single offspring are the rule, although in many species twins do occur occasionally. There is often a sharply defined breeding period so that the birth season coincides with peak rainfall, when food is abundant. The gestation period

Antelopes Family Tree

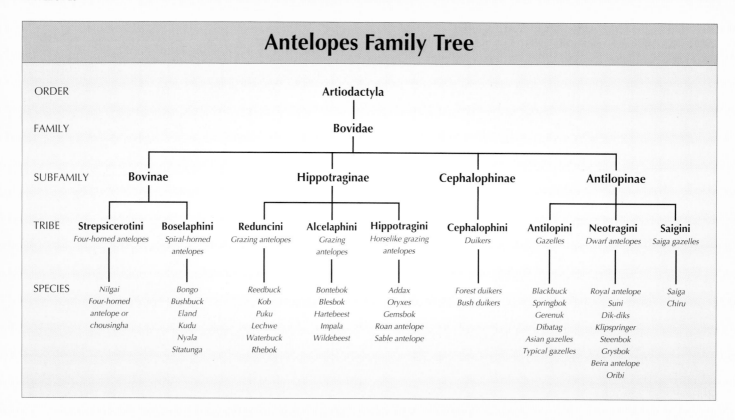

ORDER					Artiodactyla				
FAMILY					Bovidae				
SUBFAMILY	**Bovinae**		**Hippotraginae**			**Cephalophinae**	**Antilopinae**		
TRIBE	**Strepsicerotini** *Four-horned antelopes*	**Boselaphini** *Spiral-horned antelopes*	**Reduncini** *Grazing antelopes*	**Alcelaphini** *Grazing antelopes*	**Hippotragini** *Horselike grazing antelopes*	**Cephalophini** *Duikers*	**Antilopini** *Gazelles*	**Neotragini** *Dwarf antelopes*	**Saigini** *Saiga gazelles*
SPECIES	*Nilgai* *Four-horned* *antelope or* *chousingha*	*Bongo* *Bushbuck* *Eland* *Kudu* *Nyala* *Sitatunga*	*Reedbuck* *Kob* *Puku* *Lechwe* *Waterbuck* *Rhebok*	*Bontebok* *Blesbok* *Hartebeest* *Impala* *Wildebeest*	*Addax* *Oryxes* *Gemsbok* *Roan antelope* *Sable antelope*	*Forest duikers* *Bush duikers*	*Blackbuck* *Springbok* *Gerenuk* *Dibatag* *Asian gazelles* *Typical gazelles*	*Royal antelope* *Suni* *Dik-diks* *Klipspringer* *Steenbok* *Grysbok* *Beira antelope* *Oribi*	*Saiga* *Chiru*

varies between 6 and 9 months, according to species. Most species have several specialized scent glands that are important in the coordination of reproduction and social behavior.

Male breeding strategies

The size and movements of female groups define the options available for male breeding strategies. Where groups are large and range widely, for example in the eland and oryxes, genus *Oryx*, dominant males secure breeding success by following females and defending them from competitors. In other species, including hartebeests, genus *Alcelaphus*, and the waterbuck,

Kobus ellipsiprymnus, males defend territories and mate with females that come to feed there. An unusual mating system occurs in the lechwe (*K. leche*), Nile lechwe (*K. megaceros*), and kob (*K. kob*), in which males gather on special breeding arenas called leks, which females visit to choose a mate. Young males, or those in poor condition, often live in small groups (bachelor herds) that provide them with safety from predators until they challenge other males for a territory or a harem.

A small herd of Grant's gazelles, Gazella granti, *with Mount Kilimanjaro behind, Tsavo West National Park, Kenya.*

Conservation

Antelopes have been hunted by humans, for their meat and hides, since time immemorial. More recently they have also been hunted for sport. In Africa, during the last few decades of the 20th century, the bushmeat trade (the trade in wild animals as a source of food) became big business. This was largely the result of burgeoning human populations competing for ever scarcer resources in the continent. However, the biggest threat to the survival of most antelope species is competition for space with humans and their domestic livestock. The increasing demand for farmland is resulting in the loss of many of the natural habitats for antelopes. Some species, including duikers and the bushbuck, *Tragelaphus scriptus*, can adapt to these changes and visit agricultural land to feed on crops, but this puts them in conflict with farmers and ranchers.

Many species of antelopes have become threatened and some have been brought to the brink of extinction. For example, the bontebok, *Damaliscus dorcas*, was reduced to just 17 individuals in the mid-19th century before a concerted effort was made to conserve them. There are now about 1,500 bontebok in South Africa. Species with restricted ranges and specialized diets are most at risk. The most endangered antelope today is the hirola or Hunter's hartebeest, *Beatragus hunteri*, which is thought to number less than 350 in Kenya and Somalia. At least one species, the blue antelope or bluebuck, *Hippotragus leucophaeus*, has become extinct in recent times.

The future survival of several antelope species is dependent on managed breeding in either zoos or game reserves. Once captive populations have reached sustainable levels and appropriate release sites have been found, captive-bred antelopes can be reintroduced into their former ranges. Captive-breeding is being used to help save the Arabian oryx (*Oryx leucoryx*), scimitar-horned oryx (*O. dammah*), Nile lechwe and bongo (*Tragelaphus euryceros*), among others. Indeed, the scimitar-horned oryx is now thought to have become extinct in the wild. It once lived in large herds in the semiarid grassland and scrubland to the north and south of the Sahara. Small numbers of captive-bred scimitar-horned oryxes have since been released in Morocco and Tunisia.

Male sitatunga, **Tragelaphus spekei. *This unusual aquatic antelope is likely to become vulnerable in the near future.***

Antelopes are vital to the natural management of the habitats in which they live, and declining antelope populations can have serious consequences for the health of their ecosystems.

For particular species see:
- ADDAX • BEIRA ANTELOPE • BLACKBUCK
- BLESBOK • BONTEBOK • BUSHBUCK • DIBATAG
- DIK-DIK • DUIKER • ELAND • GAZELLE • GERENUK
- HARTEBEEST • IMPALA • KLIPSPRINGER • KOB
- KUDU • LECHWE • NILGAI • NYALA • ORIBI
- ORYX • PRONGHORN ANTELOPE • REEDBUCK
- SABLE ANTELOPE • SAIGA • SPRINGBOK
- STEENBOK • SUNI • WATERBUCK • WILDEBEEST

ANT LION

Pits dug by ant lion larvae under an acacia tree, Namibia. Ants and other insects fall into the pits and are seized by the larvae in their strong jaws.

ANT LION IS THE NAME given to insects of the family Myrmeleontidae, grouped in the order Neuroptera. They bear some resemblance to dragonflies and, more particularly, to lacewing flies, which they resemble in appearance and habits, in both larval and adult stages. The adults have long bodies and two pairs of long, slender wings of about equal size. Their heads are small with short antennae, knobbed at the tips. The largest species, of the genus *Palpares*, are little more than 3 inches (7.5 cm) in length, with a wingspan of up to 6⅓ inches (16 cm).

The name ant lion is appropriate because the larvae of many species feed mainly on ants. Ant lion larvae have short, thick, fleshy bodies and disproportionately large, caliperlike jaws that are armed with strong "teeth."

There are more than 1,200 known species of ant lions worldwide. A number of these are found in the United States, especially in the south and southwest, where they are called doodlebugs. Most of the North American species are around

1½ inches (4 cm) long as adults, with a wingspan of about 3¼ inches (8 cm). A fully developed larva is about ⅗ inch (1.5 cm) in length. One of the most common species in the United States is *Myrmeleon obsoletus*. The main European species are *M. formicarius* and *Euroleon nostras*.

Sandy habitats

Ant lions are found in sheltered, sandy areas such as wooded dunes, open forest floors and dry, tree-lined river banks. They also live in the sandy soil of flower beds and under hedges, the eaves of undeveloped city lots and buildings on piers. The larvae of many species burrow in the sand, the size of these burrows varying with both species and the size of the larvae. The burrow entrance is at the bottom of a conical pit, which is also dug by the larva. Groups of these pits can be readily seen in places where the soil is fine and quite dry, for example at the entrances to dry caves, beneath overhanging rocks and trees and in similar sheltered sites.

The adults are active from spring to midsummer, usually at dusk or during the night. Their flight is somewhat awkward. One reason why this type of insect is named after its larva is because the adults are inconspicuous and well camouflaged, flying only when the light is failing or has gone altogether. The larvae may also live longer than the adults.

Pit-trapped victims

Adult ant lions are reported to feed on fruits and on small flies. They might also feed on the honeydew produced by aphids, as do the lacewings, but this is unconfirmed.

The larvae of many ant lion species set traps for their prey. Buried at the bottom of its pit, with only its camouflaged head and strong jaws exposed, each larva waits for passing ants and other insects. It may flick sand to help dislodge the insect, which falls to the bottom of the pit. The larva seizes the insect as soon as it is within reach of its jaws. The prey is then drawn partially beneath the sand and the second function of the jaws comes into effect. When the maxillae or secondary mouthparts are pressed against the jaws, they together form two tubes down which a paralyzing fluid flows. This fluid is injected into the victim's body. When its struggles have ceased, digestive juices are then injected in much the same way. These dissolve the tissues, which the ant lion then sucks up and swallows. Finally, the empty case of the insect's body is tossed up and over the edge of the pit.

ANT LIONS

CLASS	**Insecta**
ORDER	**Neuroptera**
FAMILY	**Myrmeleontidae**
GENUS	***Myrmeleon, Palpares, Hesperoleon, Euroleon* and others**
SPECIES	**More than 1,200 species**

ALTERNATIVE NAME
Doodlebug (U.S. only)

LENGTH
North American species. Larva length: ⅔ in. (1.5 cm); adult length: 1½ in. (4 cm); adult wingspan: 3¼ in. (8 cm).

DISTINCTIVE FEATURES
Larva: long, protruding, caliperlike jaws. Adult: short, knobbed antennae; brown or black marks on wings; wings have many accessory veins and cross-veins; dusky body.

DIET
Larva: ants and other invertebrates. Adult: fruits and small flies; probably also some honeydew.

BREEDING
Varies with species. Number of eggs: up to 20; larva passes through 3 instars, shedding skin and digging new pit at end of each.

LIFE SPAN
Larva: up to 3 years. Adult: average 20–25 days, may be up to 45 days.

HABITAT
Sheltered, sandy areas such as wooded dunes, open forest floors and dry, tree-lined riverbanks

DISTRIBUTION
Worldwide, primarily in warm regions

STATUS
Varies according to species

Occasionally ant lion pits are grouped so close together that there is little chance of the occupants all getting sufficient food. However, it is thought that ant lion larvae are able to survive by fasting for considerable periods of time.

Not all species of ant lions dig pits. Some capture their prey by speed, and others do so by stealth or ambush, lurking beneath stones and rubbish. While the sedentary, pit-making species can walk only backwards, these more active species are able walk and move in any direction.

Long life cycle

After mating, the female ant lion lays her eggs in the sand, up to 20 in any one area. The eggs are white and oval. Being sticky on the surface, they immediately become encrusted with a layer of sand that serves as protective camouflage. Within a day of hatching, the young ant lion has already dug a pit of a size proportionate to itself. Each larva goes through three stages, known as instars. At the end of each of these, the larva leaves its pit temporarily and hides beneath the sand for about 1 week to 10 days. During this period it casts or sheds its old skin, and then digs a new pit and begins to feed again.

It is thought that the length of time spent as a larva depends to a large extent on the food available. Even with plenty of food, it is estimated that the life cycle from egg to adult takes from 1 to 3 years, longer under less favorable circumstances.

Pupation

In some species the fully grown larva pupates beneath the soil at the bottom of the pit, within a spherical, silken cocoon. Almost as soon as the silk makes contact with the air, it hardens. The cocoon is further protected by sand, which sticks to its outer surface. Only when the cocoon is completed does the ant lion larva shed its skin for the final time, revealing the cream-colored pupa. The period from pupation to emergence of the perfect winged insect depends on species.

Just prior to emergence, the ant lion pupa cuts a hole in the cocoon with its pupal mandibles and crawls part way out of the cocoon before emerging as the adult insect. At this stage, the pupal skin splits, and the adult works its way to the surface of the soil. It then climbs up a plant or tree from which it can hang while its body hardens and its wings expand and dry.

Adult ant lions live a few weeks at most and eat little, in contrast to the fiercely predatory and long-lived larvae.

ANTS AND TERMITES

ALTHOUGH THERE ARE SIMILARITIES in their form and in some aspects of their behavior, ants and termites actually belong to two different orders. The ants make up the family Formicidae and belong to the order Hymenoptera, which also includes the sawflies, bees and wasps. Termites, meanwhile, make up an entire order of their own, the Isoptera, comprising six families. There are in the region of 15,000 species of ants and these are distributed throughout most of the world. By contrast, there are fewer species of termites, somewhere in the region of 2,000, and these are mainly restricted to tropical or warm temperate parts of the globe. Termites are found as far south as South Africa, Australia and New Zealand, and as far north as Maine and Vancouver in North America, France in Europe and northern Japan in Asia.

Complex societies

Both ants and termites are known for being social insects, and live in complex societies. Both are described as being eusocial, or truly social, a major feature of which is that not all individuals are able to reproduce. In ants this responsibility falls on either a single individual (the queen) or just a few mated

Winged adults of the ant species Camponotus ligniperda. *Winged forms swarm before mating.*

CLASSIFICATION Ants	
CLASS Insecta	
ORDER Hymenoptera	
FAMILY Formicidae	
NUMBER OF SPECIES At least 15,000	

CLASSIFICATION Termites	
CLASS Insecta	
ORDER Isoptera	
FAMILIES Mastotermitidae; Kalotermitidae; Termopsidae; Hodotermitidae; Rhinotermitidae; Termitidae	
NUMBER OF SPECIES At least 2,000	

individuals (dominant ants or alpha workers), depending on species. In termites there is typically one pair of reproducing individuals in each colony, a queen and a king.

The social organization of the ants is similar to that of the termites, though there are some differences. For example, there are roughly the same numbers of males and females in termite colonies. Ant colonies, on the other hand, have many more females, the few males existing only to mate with a queen. Apart from the reproducing individuals, the other adult ants or termites exist in different forms, or castes. Each caste is adapted to perform a different task within the colony, for example building and maintaining the nest, feeding the other insects and young or protecting the colony.

The vast majority of ants and termites in any one colony belong to the sterile or unmated castes. Most of these are workers but there might also be a smaller number of soldiers. The workers build the nest around the royal chamber. In addition, they provide food and tend to the next generation of young. Soldiers, meanwhile, possess mouthparts adapted for attacking predators, and may even possess structures that propel unpleasant fluids at their attackers. Some ants have evolved a true sting. In comparison to termites, few ant species have a soldier caste, however. Where they do it is more a case of some workers adopting the soldier functions and defending the colony against predators, which may include other species of ants.

Ants have a "waist"

The main difference in appearance between ants and termites is that ants have an obvious "waist" formed in their abdomen, while termites do not. Despite this difference, termites are sometimes known as white ants, because of their pale, soft abdomen. The abdomen of both ants and termites is variable in form, and reaches a specialized form in honey ants. The honey ant is not a particular species, but a form of worker, called a replete, found only in some species. These individuals store a sweet solution in their enlarged bodies, and provide it to other ants within the nest. So nutritious is this fluid, some humans are known to seek out these ants to drink their "honey."

Termites and ants both have winged, dispersing forms that emerge periodically to swarm and mate. The timing of the swarm and the abundance of these individuals depends on many factors, including climate.

Different life cycles

Despite the similarities between ants and termites, their life cycles are quite different. Both insects start as eggs, but the young termites resemble their parents in basic body structure. They change only a little as they develop through their imma-

Termite mounds, such as this impressive 20-foot (6-m) example in Botswana, are called termitaria. They are made of sand or soil mixed with the termites' saliva or excrement.

ture nymph stages and molts to become adults. In termites, some immatures will even assume worker or soldier duties around the nest. This compares to the young ants, or larvae, which bear little resemblance to adult ants. They undergo a complete change of form from larva to adult, going through a pupal stage in which most of the transformation occurs.

Termite mounds and ant nests

Termites are famed for their ability to form large mounds, called termitaria. The mounds are made from sand or soil mixed with saliva and excrement. This mixture forms a very strong structure, so much so that the Australian termite, *Nasutitermes triodiae*, is able to construct a mound more than 23 feet (7 m) high. Nests can be domelike, wedge-shaped or conical. Within each mound there is a complex network of passages and chambers maintained by the workers.

There are many termite species that do not create mounds, however, and instead live in nests that are little more than a series of passages within wood. The wood-feeding habits of some species, like those of the genus *Reticulitermes*, make them particularly damaging to woodwork. Other species, such as *Neotermis militaris*, form underground nests or feed inside growing plants, sometimes causing damage to crops. Termites are often considered pests, although only 10 percent of species are damaging either to wood or commercial crops. Ants also form nests above and below ground, the size and design varying from the very simple to the large and complex series of chambers and galleries found in nests of the leafcutter ants. Some ant nests are very large: those of the wood ant, *Formica rufa*, may reach 6½ feet (2 m) below ground and are topped with mounds of leaf litter up to 5¼ feet (1.6 m) high.

What do they eat?

The mouthparts of ants and termites are generally adapted for chewing, though in some cases they are specialized more for tasks such as defending the nest. Both ants and termites eat a variety of foods. Some are predatory, while others, such as the leafcutter ants, have adopted a farming lifestyle. In this they collect fragments of leaves and place them in special "garden" chambers within the nest. A fungus grows over the leaves in the chamber, and the ants feed on the fungus. The ants may even use natural antibiotics to keep unwanted bacteria and fungus out of their gardens.

Other ant species drink a sweet fluid called honeydew, excreted by other insects such as aphids, psyllids, scale insects and some butterflies and moths. The ants, collectively known as pastoral ants, may actually "milk" these other insects of their honeydew by stimulating them with their antennae. Ants will also protect these insects from predators and maintain their food source in other ways, for example by herding them to the most nutritious parts of a plant. Some ants even take the insects or their eggs back to their nest.

Worker termites of the genus Nasutitermes *building their nest, Rakata Islands, Sunda Straits, Indonesia. When complete, the nest will contain a complex network of passages and chambers.*

Worker termites dragging dry grass stems back to their nest. They are guarded by soldiers (top left of photograph) as they work.

Termites feed on cellulose, so wood is an important food source for them. The gut of wood-eating species contains microbes that help them to break down and digest the tough structures. Some termites do not eat wood firsthand, but instead eat the semidigested material that has already passed through the guts of other individuals in their colony.

Founding the colony and reproducing

Whereas a reproducing king and queen head termite colonies, in ants there is only a queen or several queens because the males die soon after mating. Queen ants may join existing colonies rather than start new ones, but generally ant and termite colonies are founded in a similar way, by swarming.

In termites, winged, dispersing individuals called alates develop at a certain time of year from the primary reproducer caste. When conditions are favorable, they exit the nest through tunnels prepared by the workers and, being weak fliers, are carried by the wind. Shortly after this flight, they shed their wings and mate with a member of the opposite sex. A new termite colony is founded by the pair. They form a chamber around which the nest will be built. In some species the queen can reach 4 inches (10 cm) in length, her body enlarged with eggs and stored food. The king is normally just

⅖–⅘ inch (1–2 cm) long. Another caste, the supplementary reproducers, may also be present in a termite colony and can replace the original king or queen if they die.

In ants, the colony produces a host of new, winged queens, also at certain times of the year when conditions are favorable. The queens leave the nest together to embark on so-called nuptial flights, during which they will mate with winged males, before founding colonies of their own.

Social parasites and slave-makers

Females of the ant *Anergates atratulus* lay their eggs within the nest of another ant species, *Tetramorium caespitum*. The young *Anergates* are then tended by the *Tetramorium* workers, and eventually give rise to a whole generation of invaders. In some cases, the queen of the host colony will be killed and the host species will die out completely, leaving the nest to the invaders.

Slave-making ants, such as *Polyergus rufescens*, actually kidnap pupae from the nests of other species. These pupae are taken to the slavers' nest, where they emerge as adults and work for their captors. Some slave-maker ants have completely lost the ability to produce workers of their own.

For particular species see:
• ARMY ANT • HARVESTING ANT • HONEY ANT
• LEAFCUTTER ANT • PASTORAL ANT • TERMITE
• WOOD ANT

APES

EVER SINCE THE ENGLISH naturalist Charles Darwin suggested that humans descended from apes, the family of this animal group has been hotly debated. Today, although few scientists dispute that *Homo sapiens* had its origins in apelike forms, subdivisions within this animal group continue to be redefined regularly. A currently popular family tree places all the apes in the superfamily Hominoidea. This is then broken into three families, comprising 11 species of lesser apes or gibbons in the genus *Hylobates* and family Hylobatidae;

the great apes family, Pongidae, which includes the orangutan, chimpanzee, bonobo (pygmy chimp) and gorilla; and finally humans, the sole member of the family Hominidae.

Origins

The apes probably separated from the monkeys around 30–25 million years ago in what is now Africa. Between 20–15 million years ago they developed into many species, including the gibbons, and reached Asia. The major split between the orangutan, *Pongo pygmaeus*, and the other great apes probably occurred about 12.5 million years ago, with the African hominids, chimps and gorillas diverging as recently as 6 million years ago. Several bipedal (walking on two legs) hominid species evolved and became extinct during the few million years leading up to the appearance of *Homo sapiens*, just 1.5 million years ago. The evidence for the close link between *Homo sapiens* and other apes is now overwhelming. We share, for example, 98.4 percent of our genome (genetic

*The bonobo, **Pan paniscus**, resembles the chimpanzee, only it is more slender, with a black face and a smaller, rounder head.*

CLASSIFICATION
CLASS Mammalia
ORDER Primates
SUPERFAMILY Hominoidea
FAMILY Hominidae: humans; Pongidae: great apes; Hylobatidae: lesser apes
NUMBER OF SPECIES 15

material) with the chimpanzee; not even horses and zebras are so closely related.

Physical characteristics

Compared to monkeys and other primates, apes typically possess large, tail-less bodies with broad, barrel-like chests. The forelimbs are long, spectacularly so in the gibbons. The brain is large, complex and capable of advanced thinking. Common to all apes, including ourselves, is a dental formula of 32 teeth: two incisors, one canine, two premolars and three molars (eight teeth) on each side of both the upper and lower jaws (multiplied by four).

However, there are key differences among the apes. Gibbons have a slighter build and are smaller, with a head-and-body length of about 30–36 inches (75–90 cm). Their fingers are long and slender, and the thumb joint is set back almost to the wrist. The buttocks bear toughened pads. Different species of gibbons show a tremendously rich range of color variation, even between sexes of the same species.

Much bigger and heavier, the great apes are very muscular and have larger brains than the gibbons. A male eastern lowland gorilla, *Gorilla gorilla graueri*, can top 385 pounds (175 kg), more than twice the weight of an adult human.

Life among the leaves

Humans have all but overrun the globe, but wild apes are essentially restricted to forested regions. Gibbons live in the deciduous and evergreen rain forests of tropical Asia from India east to Southeast Asia and China. Active by day, they cover as much as 1 mile (1.6 km) each day through the forest canopy, up to 130 feet (40 m) above the ground. Gibbons are experts in brachiation: swinging arm-over-arm beneath branches, forming hooks with their hands.

The gorilla, chimp (*Pan troglodytes*) and bonobo (*P. paniscus*) are found in the forests of equatorial Africa, with the chimp appearing also on wooded savanna. While they are far less arboreal (tree-living), these great apes do nest at night among the branches. They tend to walk quadrupedally (on all

The white-handed gibbon, Hylobates lar, *has a distinctive singing voice, which it uses to proclaim its territory and develop bonds with other gibbons. Young gibbons sing solo until they find a mate, at which point they begin to sing together in duet.*

fours), supporting the forequarters on their knuckles, although like the gibbons they are capable of upright, bipedal locomotion. The orangutan lives only on the tropical islands of Borneo and Sumatra. It is more arboreal than the other great apes: a cautious climber, this bulky primate keeps a firm grip with its hands and feet, yet readily ventures up to 100 feet (30 m) into the leafy canopy.

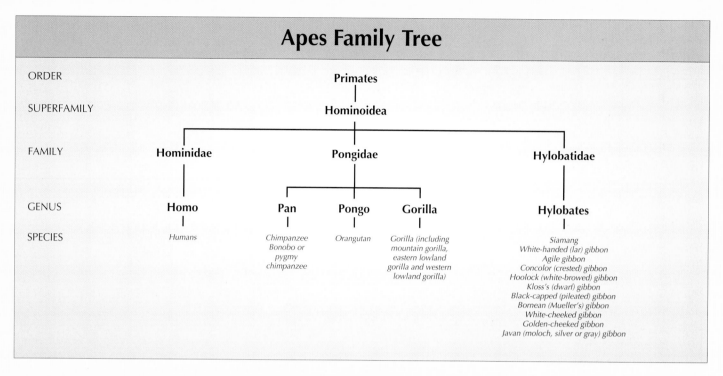

Apes Family Tree

ORDER			Primates		
SUPERFAMILY			Hominoidea		
FAMILY	Hominidae		Pongidae		Hylobatidae
GENUS	Homo	Pan	Pongo	Gorilla	Hylobates
SPECIES	*Humans*	*Chimpanzee* *Bonobo or pygmy chimpanzee*	*Orangutan*	*Gorilla (including mountain gorilla, eastern lowland gorilla and western lowland gorilla)*	*Siamang* *White-handed (lar) gibbon* *Agile gibbon* *Concolor (crested) gibbon* *Hoolock (white-browed) gibbon* *Kloss's (dwarf) gibbon* *Black-capped (pileated) gibbon* *Bornean (Mueller's) gibbon* *White-cheeked gibbon* *Golden-cheeked gibbon* *Javan (moloch, silver or gray) gibbon*

Versatile eaters

Leaves, fruits, other plant parts and occasionally eggs or live prey are typical ape food, but within these limitations many species have a bewilderingly varied diet. The orangutan uses more than 400 plants, gorillas more than 100 and chimps at least 250. Chimps also gang up to hunt down and kill vertebrates, including antelopes and rodents.

One of the more impressive eating habits of apes concerns their use of tools to extract favorite foods. The chimpanzee takes a moistened twig or grass stem to winkle termites from their mounds. It also sponges up drinking water by using bunched leaves and picks up heavy sticks or stones for use as weapons or as anvils to crack nuts. Just as infants learn what plants are good to eat by watching their mother, so they can develop the skills of using tools. However, as in humans, it can take several years of training to perfect such techniques.

A bonobo clutches its feet, showing the enlarged toe, or "thumb," opposing the four other toes. This increases the bonobo's agility in climbing and manipulating objects.

sexual "badge." At the other extreme, the chimpanzee forms large, male-dominated groups of several adult males and females that engage in dynamic interaction. Gorilla groups are ruled by massive, mature adult males known as silverbacks because of the silver-gray color of their coats.

Gibbons, like other apes, are highly territorial. Adult pairs reinforce their claim to their portion of land each day by singing deafeningly loud, exotically modulated duets that can carry several miles through the trees. These calls vary widely across the different species. Great apes, also highly vocal, are known particularly for their range of facial expressions to communicate mood.

Breeding

Apes are very slow breeders. Usually a single offspring is born, and typical birth intervals range from 30 months in the white-handed or lar gibbon to 8 years in the orangutan. This extended pause between breeding stems from the extremely long period of parental care (up to 3 years in great apes) that is required to prepare such advanced brains for a complex lifestyle. However, as is usual with animals that have a slow rate of reproduction, the great apes are all long-lived. The gorilla lives for up to 50 years in the wild, and wild chimpanzees may live to be 60 years old.

A baby orangutan holds on tight to its mother's long hairs as she dangles from liana vines in the rain forest. For the first year of their lives, baby orangutans depend solely on their mothers to provide them with food and teach them the skills to survive.

Many ape populations are in deep trouble, not least as a result of continuing deforestation throughout tropical woodlands. Orangutan numbers were devastated by raging forest fires in Southeast Asia during 1997. Civil unrest in several African countries, along with hunting for meat, continues to threaten African great apes, most notably the mountain gorillas, *Gorilla gorilla beringei*, of the Rift Valley's Parc des Volcans. Recent protection efforts include a proposal from New Zealand to bestow full human rights on humanity's closest cousins, the great apes.

Advanced societies

Apes demonstrate a range of social structures, but the most prevalent are family groups led by a mated pair and containing juvenile offspring of one or more generations. In the orangutan, the female and young live apart from other females and their offspring and away from the solitary adult males. The massive cheek flanges of adult males serve as a

For particular species see:
• CHIMPANZEE • GIBBON • GORILLA • ORANGUTAN

APHID

APHIDS OR PLANT LICE are a group of hemipteran insects of great economic importance because they do such harm to crops. Damage is done both directly, by the aphids sucking sap from plants, and indirectly, by their transmitting certain diseases from one plant to another. There are about 4,000 species of aphids, including the greenflies and blackflies, which are not true flies. They are probably the most important pests in temperate regions, where one plant species in four is infested. Aphids have soft, oval bodies, small heads, compound eyes, long antennae with three or six joints and a jointed beak, or rostrum, adapted for piercing plant tissues. Some have transparent wings. They range in size from ⅕ to ⅖ inch (1–10 mm).

Migrate on air currents

Most aphid species remain on a single host plant and several or all generations comprise parthenogenetic females, which reproduce without mating. Other aphid species move from a primary winter host plant to a secondary spring and summer host. For example, after mating in late summer or the fall, the black bean aphid, *Aphis fabae*, lays eggs on the spindle tree or guelder rose. These hatch the following spring as winged females, which fly to bean crops where they reproduce parthenogenetically. The peach-potato aphid, *Myzus persicae*, also follows this type of life cycle.

When migrating between plants, the winged females are carried up on air currents, often to a great height. After several hours, descending air currents bring them down and they seek out suitable plants. The aphids may be carried for hundreds of miles in this way.

Piercing and sucking feeders

The vast majority of aphid species are extremely host specific, feeding on one or a few related plant species. The mouthparts are modified for piercing plant tissues and sucking up the cell-sap, especially from the phloem, the main food stream of the plant. Before feeding begins, a salivary secretion is injected into the wound to prevent the sap from coagulating as it flows.

While many aphids feed externally on plants, others form galls, or enclosed receptacles, in which they are able to feed while hidden from predators. Examples of aphid gall-makers may be found on trees such as poplars, elms, limes and spruce and on cultivated currant bushes. Frequently the galls form a refuge for passing the winter, for instance in the woolly aphid or American blight, *Erisoma lanigerum*, of apple trees, which makes a fluffy "wool" in which it feeds.

Reared by ants

Probably the most striking fact about aphids is their excretory habit. When feeding, aphids take up large quantities of sap in order to get sufficient protein. The rest, a fluid rich in sugar, is given out through the anus as honeydew. Being rich in sugar, honeydew is much sought after by ants and some other insects. Ants rear, or at least closely associate with aphids, eating the honeydew and taking it back to the ant larvae. They encourage the aphids to produce honeydew by improving conditions for their existence. For example, they may herd their charges to the growing tips of plants, which are the most nutritious, thus stimulating growth and the emission of honeydew. Where there are no ants present, the honeydew may eventually cover large areas of the plant, causing its death by suffocation or by attracting fungi. By removing the honeydew, the ants ensure the aphids' food sources.

Ants also take aphids into their nests, where the aphids may lay eggs. Alternatively, the ants may carry the aphid eggs from the plants on which they are laid back to their nest. After emerging, the young aphids are carefully tended and "milked" by the ants while they feed on the roots of various plants.

Female aphid giving birth to young. All aphid species are able to reproduce parthenogenetically, that is, without mating.

APHIDS

PHYLUM	**Arthropoda**
CLASS	**Insecta**
ORDER	**Hemiptera**
SUBORDER	**Homoptera**
SUBFAMILY	**Aphididae**
GENUS	**Many, including *Acyrthosiphon*, *Aphis* and *Microlophium***
SPECIES	**About 4,000 species**

ALTERNATIVE NAMES
Plant louse; greenfly; blackfly; ant cow

LENGTH
⅟₂₅–⅖ in. (1–10 mm)

DISTINCTIVE FEATURES
Small, soft body; mouthparts extended to form tube for sucking sap from plants; 2 siphunculi (tubes) extend from abdomen

DIET
Phloem sap sucked from plants

BREEDING
All species breed parthenogenetically (without mating); produce many young when 8–10 days old. Some species also reproduce sexually; eggs hatch following spring.

LIFE SPAN
3 weeks to 1 month

HABITAT
Anywhere with suitable host plants

DISTRIBUTION
Worldwide, especially in temperate regions

STATUS
Superabundant

Young without mating

For most of the year, aphid populations consist only of females, which reproduce parthenogenetically at a great rate. All aphid species are able to reproduce in this way.

Some species also reproduce sexually at certain times of the year (usually in the fall), and eggs are laid which hatch the following spring. In this case, winged females fly back to their primary host plants (usually trees) and lay eggs, which hatch as both males and females. These mate and lay eggs in crevices in the bark. When the new eggs hatch, they produce only females, and thus the cycle begins over again.

A single parthenogenetic female may produce as many as 25 daughters in one day, and as these themselves are able to breed when 8 to 10 days old, the numbers of aphids produced by just one female in a season can reach astronomical proportions. Breeding is slowed down, however, by adverse conditions, especially cold.

Black bean aphids feeding on plant sap. Many aphid species do great harm to commercial crops.

Predators everywhere

Many small insect-eating birds, such as tits and flycatchers, eat aphids. Ladybugs, lacewings, bugs, spiders and hoverfly larvae also prey on them. In addition, certain parasitic wasps of the family Braconidae lay their eggs in aphids. The wasp larvae consume the tissues of their hosts and eventually pupate inside the empty husks.

Some aphids, however, have evolved defensive mechanisms to guard against attack from these wasps. The aphids' blood cells secrete a capsule that envelops the parasite larva, arresting its development completely. Aphids are also able to deter insect predators by exuding a kind of wax from a pair of chimneylike stumps, called cornicles, on their rear end. This temporarily paralyzes the attacker.

Any ants that are rearing aphids for their honeydew will also protect their charges from attack by predators.

APOLLO BUTTERFLY

COMMON APOLLO

PHYLUM	**Arthropoda**
CLASS	**Insecta**
ORDER	**Lepidoptera**
FAMILY	**Papilionidae**
GENUS AND SPECIES	***Parnassius apollo***

T HE NAME APOLLO BUTTERFLY is used for any of 30 or more species of the genus *Parnassius*, part of the family of butterflies known as the swallowtails. In North America, Apollo butterflies are also known as parnassians.

The Apollo butterflies are not brightly colored, most of them being white, with spots and eyelike markings of black and red. They are unusual in their flight, sometimes soaring on rising air currents, wings outstretched.

Range and habitat

The common or European Apollo, *P. apollo*, is found in mountainous regions of Europe, from Scandinavia to the Alps and Pyrenees, but not in the British Isles. It flies at relatively low altitudes. A related species, the alpine Apollo, *P. phoebus*, occurs at higher altitudes. Although about 30 species of Apollo butterflies are known, ranging through Europe and Asia to western North America, the taxonomy of *Parnassius* is changing and there may be more than this. All known species are mountain butterflies, some ranging up to 20,000 feet (6,100 m) in the Himalayas.

Life history

The caterpillar of the common Apollo feeds on orpine, a species of stonecrop, in addition to the houseleek. The larvae of other species feed on a variety of hosts, including plants of the families Fumarialeae, Scrophularaceae and Crassulaceae.

The common Apollo larva is black with red spots and, when fully grown, spins a cocoon in which to pupate. The caterpillar of one of the North American species, *P. autocrator*, is brilliant orange in color and gives off a pungent odor when threatened.

The common Apollo is now recognized as being an endangered species, the result of collection by lepidopterists, pollution and habitat destruction.

ALTERNATIVE NAME
European Apollo

LENGTH
Adult wingspan: 2–4 in. (5–10 cm)

DISTINCTIVE FEATURES
Larva: black with red spots; spins a cocoon in which to pupate. Adult: white with spots and eyelike markings of black and red.

DIET
Larva: plants such as orpine and houseleek. Adult: flower nectar.

BREEDING
Breeding season: May–September

LIFE SPAN
Varies with temperature and environment

HABITAT
Mountainous regions at altitudes of 1,640–7,900 ft. (500–2,400 m)

DISTRIBUTION
Europe, from Scandinavia to the Alps and Pyrenees Mountains, and east across Asia

STATUS
Endangered

Common Apollo

The duration of the life cycle depends on temperature and environment. In some cases Apollo butterflies can develop fully in 1 year, others require 2 years. The habit, very unusual among butterflies, of spinning a cocoon is perhaps related to the need for protection from frost at high altitudes. However, the cocoon might be a defense against egg-laying by parasitic wasps.

ARCHERFISH

ARCHERFISH ARE ANY OF six species of fish, which reach up to 10 inches (25 cm) in length. They are known for their ability to obtain insects by shooting them down with a stream of water drops. Archerfish can be found from India, through Southeast Asia and the Malay archipelago, to parts of Australia and east as far as the Philippines. Their habitat is mainly the salty waters of mangrove belts, but they may also live in the sea, or go up streams into fresh water.

Precision water pistol

The main food of the archerfish consists of small water animals swimming or floating near the surface, and insects crawling on the leaves and stems of overhanging vegetation. A fully grown adult can hit insects 6 feet (1.8 m) above the surface of the water. Indeed, one archerfish was seen to miss its target, but the jet of water traveled a measured 15 feet (4.6 m). If the fish misses with its first jet, it will follow with several more in rapid succession. At the moment of shooting, the tip of the snout breaks the water, but the eyes are submerged. Water in the gill chambers is driven into the mouth by a sudden, powerful compression of the gill covers. At the same time the tongue is pressed upwards, converting a groove in the roof of the mouth into a tube, which increases the speed of the outgoing stream.

Archerfish begin to spit when very young and only 1 inch (2.5 cm) long, but the jets of water they produce do not travel much over 4 inches (10 cm). As they grow, their aim and the length to which they are able to spit improves.

Iridescent young

Adult archerfish apparently spawn far from land, in coral rock or reefs. The young return to the mangrove belts, or into freshwater rivers. Like the adults, they have dark bars on their backs, but they also have "light-flecks," yellow iridescent flecks, between the dark bars. At times the flecks shine so brightly they appear like tiny, green fluorescent lights. It is suggested that they may act as recognition marks between young members of the species, helping them remain together in the muddy waters.

Aiming from underwater

For a long time it was not known how an archerfish was able to judge distance and take aim at targets in the air above while its eyes were below water. For example, if a stick is dipped at an angle into water, it appears bent due to the bending of light rays as they pass from water

into air. This is known as refraction. The archerfish would therefore see its food in a different position from the true one, yet it still managed to have a very good aim. It was assumed that the archerfish must allow for refraction. However, the truth was revealed in 1961 by a more careful observation of its feeding techniques. The archerfish swims forward until it is almost under its target and appears to take aim. Then, as it ejects the water jet, it tends to jerk its body nearly to the vertical. In this position, just as a stick dipped straight into the water does not appear to bend because refraction is reduced to a minimum, the archerfish, by looking straight up out of the water, sees the exact position of its insect target.

It has been noted that archerfish sometimes miss their target. This is probably when, over-eager to take aim, they shoot their water drops before they have positioned themselves as nearly vertical as possible. If they miss, they try again. Their marksmanship improves with practice, which indicates learning ability.

A spotted archerfish, Toxotes chatereus, spits a stream of water drops at a spider sitting on a leaf overhanging the water. If the aim of the archerfish is true, the spider will be shot down into the water, providing the fish with its next meal.

Archerfish can also leap up out of the water and catch prey in their mouths, as in this action shot of an archerfish pursuing an insect it has spotted above the surface.

BANDED ARCHERFISH

CLASS	Osteichthyes
ORDER	Perciformes
FAMILY	Toxotidae
GENUS AND SPECIES	*Toxotes jaculator*

LENGTH
Up to 10 in. (25 cm)

DISTINCTIVE FEATURES
Large eyes; strongly pointed snout; variable coloration and markings. Adult: usually silver overall with greenish tinge to certain areas; 4 to 6 broad, black transverse bars that become shorter with age; yellowish green dorsal fin; dirty green tail fin; silver anal fin with broad black margin. Juvenile: yellowish green to brown above; grayish green flanks; silvery white belly.

DIET
Small aquatic animals near water's surface; also insects and other invertebrates on overhanging leaves and stems

BREEDING
Poorly known. Age at first breeding: when about 4 in. (10 cm) long.

LIFE SPAN
Not known

HABITAT
Mainly in salt water in mangrove swamps and at river mouths; also upstream in fresh water and out to sea

DISTRIBUTION
Coastal regions in Indian and South Pacific Oceans, including southern Middle East, India, Malay archipelago, Philippines, Australia and South Pacific island groups

STATUS
Not known

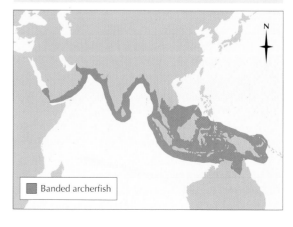

Banded archerfish

Curiouser and curiouser

The eyes of archerfish are larger and more highly organized than in most fish, giving binocular vision, meaning the eyes are forward facing, with a large overlap of vision. This allows the fish to focus well and judge distances accurately. Other specializations of the archerfish include its mouth and tongue shape, the mechanism for producing a jet of water, and the behavioral adaptations associated with these. When an animal has many adaptations to achieve one end, it is assumed these have been brought about by the pressure of natural selection and are therefore necessary for its survival. Yet in spite of its elaborate equipment for shooting down insects, archerfish do not use this as the main way to obtain food. They can get along quite well without using it. Archerfish that live in the sea apparently never shoot insects down.

Perhaps even more odd is the way archerfish will use their jets unnecessarily. They have been seen to aim their jets of water at insects already fallen on the surface, which they could easily have snapped up directly. In addition, they have been seen to direct a jet at a small object, edible or inedible, lying on the bottom. Perhaps the archerfish instinctively shoots at any interesting target and uses these opportunities as practice to impove its accuracy. Archerfish will also catch insects flying low over the water by leaping up and seizing them with the mouth.

ARCTIC AND ANTARCTIC

THE ARCTIC AND THE ANTARCTIC, situated at the North and South Poles respectively, are the two coldest places on earth. Within these regions there is both a great abundance and a great absence of life. This wildlife is fascinating in its adaptation to the extremes of climate and habitat found in these hostile environments.

The polar regions do not sit easily inside the Arctic and Antarctic circles that we see on maps. These circles are defined by the lines of latitude within which, for one day or more a year, the sun does not set. "Midnight sun," as the phenomenon is called, occurs during the summer or winter solstices, usually on or around June 22 and December 22. The Antarctic lies below the line of latitude at 66° 30′ South, while the Arctic lies above 66° 30′ North. However, this is an artificial definition. A true physical definition of the Arctic and Antarctic as *polar* is difficult because of the very different terrain of the two poles.

The Arctic is a largely oceanic area, covered by sea ice, this being surrounded by land. Tundra, that is treeless regions with a permanently frozen subsurface of black, peaty soil (see article under "Tundra"), covers much of this landmass. The Antarctic, on the other hand, is a distinct continent that extends outside the Antarctic circle and is surrounded by the Southern Ocean. The polar region here covers the continental ocean out to the so-called Antarctic convergence, where the cold polar oceans meet the warmer South Pacific, South Atlantic and southern Indian Oceans.

History of the polar regions

The low angle at which the sun strikes the poles reduces the heat available to warm the earth's surface in the polar regions. This results in a cold climate and ice formation.

In the last ice age, some 25,000 years ago, the Arctic ice cap was much larger. At this time the ice "locked up" so much of the world's sea water that the sea level was some 300 feet (90 m) below what it is today. The weight of the ice forced the earth's crust down, forming the present Arctic Ocean. The ice cap also tilted the earth's axis of rotation, warming the pole and providing a self-regulation of the global climate.

Grounded icebergs in the Antarctic. Whereas the Antarctic is a continent capped with ice, the Arctic is largely a frozen sea.

Some 200 million years ago the landmass that now forms Antarctica was in the same place it is now. However, it is thought that it was part of the much larger Gondwanaland supercontinent at that time. Theoretically, continental drift moved the landmass northward and broke it up so that, by 70 million years ago, the climate was semitropical. By 60 million years ago, Antarctica and Australia are thought to have broken apart. Antarctica then moved south, so that by 25 million years ago ice was replacing the forests.

The Antarctic ice cap at present contains 90–95 percent of the world's freshwater ice. The volume of Antarctic ice is estimated at 7 million cubic miles (30 million cu km). Its weight depresses the bedrock beneath it so that much of the continent is below sea level, with the lowest point being 8,325 feet (2,540 m) down. The Arctic ice cap, by contrast, is salt water.

One of the definitive Antarctic birds is the Adélie penguin, **Pygoscelis adeliae,** *which lives in the frozen seas that surround the continent. Most other penguins live farther north, in warmer subantarctic waters.*

Climate

Both the North and South Poles have a cold and very dry climate, and can be considered ice deserts. The slight tilting of the earth's axis relative to the sun leads to extreme seasons at the poles, with 6 months of daylight followed by 6 months of darkness. The Antarctic summer is in the Arctic winter and vice versa. The long winter only emphasizes the cold. The Arctic can reach -106° F (-77° C), but the Antarctic is the coldest place on earth, with a record -126° F (-88° C).

The Antarctic has the harsher climate because it is also more windy, with average annual wind speeds around some parts of the coast being 45 miles per hour (72 km/h). In the Arctic the ice covers and insulates the ocean which, as it never freezes, maintains a higher temperature than in the Antarctic. Average annual temperatures in the Antarctic never exceed 14° F (-10° C), a level that

ARCTIC AND ANTARCTIC

CLIMATE
Very arid, with extreme seasonal variation in day length. Annual rainfall: 1¼–3 in. (3–7 cm). Maximum temperature: 68° F (20° C) in summer sun. Minimum temperature: -106° F (-77° C) in Arctic; -126° F (-88° C) in Antarctic.

VEGETATION
Antarctic landmass: algae, fungi and lichens. Seas and oceans (both poles): phytoplankton.

LOCATION
Arctic: area north of treeline of North America and Siberia. Antarctic: a distinct continent, including subantarctic islands such as South Shetlands and South Orkneys; Antarctic Ocean is defined by sharp change in temperature (Antarctic convergence).

STATUS
Global warming, due in part to depletion of the ozone layer (particularly over the ice caps), is reducing the extent of ice cover and causing the break-up of more ice sheets. This is leading to a global rise in sea levels.

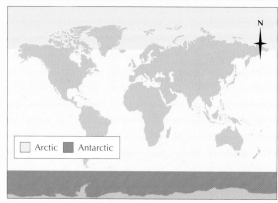

Arctic Antarctic

the islands of Svalbard in the Arctic may reach only in January. The low elevation of the Arctic also helps to keep wind speeds down.

The Antarctic has a high ice mass and even mountain ranges. Much of the center of the continent is more than 6,560 feet (2,000 m) high. This high altitude causes gravity, or katabatic, winds, in which air in the cold, high center becomes dense and is pulled down to the warm coast. The Antarctic ice also loses more heat than it absorbs from the sun. Were it not for a flow of warmer, wetter air from the world's temperate regions, the continent would become colder and colder. The average snowfall in this region is the equivalent of only 1¼–3 inches (3–7.5 cm) of rain. The accumulation of up to 15,670 feet (4,775 m) of ice has taken millions of years.

Habitats at the Poles

In both the Arctic and the Antarctic, most life depends on the relative warmth and stability of the ocean habitat. In the Antarctic the ocean remains close to the freezing point of salt water; the sea is 23° to 30° F (-5° to -1° C), depending on its salinity. This, and the supply of nutrients brought by the ocean's currents that upwell where they meet the Antarctic ice cap, provides a

superabundance of microscopic phytoplankton (simple plants). These plants form the basis of the food chain. The Arctic is a relatively nutrient-poor area, by comparison. This is because its stable currents are contained within the surrounding landmass. In both polar regions, the marine ecosystems are based, unusually, around one group, the small crustaceans known as krill. Krill provide the food for whales, seals, fish and birds.

There is no land under the Arctic ice cap, although Arctic ice floes provide shelter and a fishing platform for various mammals and birds. The Antarctic, by contrast, provides both terrestrial and freshwater habitats, although they are inhospitable. The ice covers all but 2–3 percent of the continent. The ice-free valleys, or oases, are strange, self-perpetuating habitats. Sufficient heat is absorbed by the rocks to keep the occasional snowfalls away. These cold deserts, averaging temperatures of -4° F (-20° C), mean that life there is limited to microorganisms. Even these organisms are virtually confined to the glacier-melt streams or the saline lakes that the streams feed. The lakes are brackish enough at the surface to support life, but evaporation into the extremely dry air forms heavier salts, which sink to create a dense layer of salt water at the bottom.

Pack ice breaking up at Foxe Basin, Canada. Melting polar ice is leading to rising sea levels worldwide.

Seals are found on and beneath the sea ice of both the North and South Poles. Pictured is a bearded seal, Erignathus barbatus, in the Arctic Ocean.

Polar plants

The dominant plants at both poles are the single-celled diatoms that form the majority of the photosynthesizing plankton. These are the basis of the marine food web.

On the Antarctic continent green algae are able to survive in the snowfields. Despite their name, these algae can be pigmented green, red or yellow and often color the glaciers and ice caps. In the ice-free areas, the lack of soil does not prevent some species of algae, fungi and lichens from surviving. Lichens predominate because of their ability to grow on the bare rock and to freeze dry in the winter and then rehydrate in summer. Only two flowering plants are known in Antarctica, the Antarctic hair grass, *Dechampsia antarctica*, and the Antarctic pearlwort, *Colobenthos sublatus*. However, these plants are found only on the west coast of the Antarctic peninsula, where peaty soils have accumulated in the wetter areas.

In the Arctic, outside of the southern tundra areas, only subsurface algae, called the epontic community, survive on the ice floes. The Arctic is devoid of other plants.

Polar animals

The larger animals that most people think of as being typically polar are dependent on the richness of the marine ecosystems at the poles. They include species such as seals, dolphins, whales, penguins and other seabirds. However, the isolation and far harsher climate of the Antarctic produces a radically different terrestrial animal population from that of the Arctic. The largest

land predator in the Arctic, the polar bear, *Ursus maritimus*, may weigh up to 1,720 pounds (780 kg). The largest in the Antarctic, a species of mite, *Gamasellus racovitzai*, weighs just ¹⁄₂₅₀ ounce (0.1 g). Other invertebrates living in the Antarctic include springtails, ticks and wingless flies.

Although the Antarctic and subantarctic support a variety of albatrosses, petrels, cormorants, terns and skuas, the definitive Antarctic bird group is the penguins (family Spheniscidae). The extreme specialization of penguins is possible only because of the virtual absence of predators on their breeding grounds. The largest species of penguin, the emperor penguin, *Aptenodytes forsteri*, breeds during the perpetual darkness of the Antarctic winter and endures colder breeding conditions than any other bird. It is the male emperor penguins that incubate the single eggs. During blizzards, when the air temperature may plummet to -70° F (-55° C), the male penguins gather together, each resting its bill on the bird in front. In this way they reduce heat loss by 25–50 percent.

The sea ice of both poles provides a habitat for seals. The ringed seal, *Phoca hispida*, of the Arctic, spends the winter beneath the protecting ice and uses blow holes in the thinner ice to breathe. One of the most common seals of the south, the Weddell seal, *Leptonychotes weddellii*, may have no polar bears to deal with, but is prey to killer whales or orcas, *Orcinus orca*, and leopard seals, *Hydrurga leptonyx*. Crabeater seals, *Lobodon carcinophagus*, are the most numerous seals in the Antarctic, if not the world. Their population may be up to 15 million.

All of the mammals of the poles have had to adapt to the cold temperatures. Typically such adaptations are thick fur and insulating blubber, or fat, beneath the skin. Although it appears to be white, polar bear fur consists of hollow, colorless hairs. These conduct heat to the polar bear's skin, which is black and therefore a highly efficient heat absorber. A thick layer of fat under the skin, 3 inches (7.5 cm) thick on the haunches, insulates the polar bear and keeps it buoyant in the water.

Global warming

The main threat facing polar biomes is global warming. Due in part to the depletion of the ozone layer, this phenomenon is greatly reducing the extent of polar ice cover and is causing the break-up of more ice sheets. This is leading to a global rise in sea levels. Recent studies suggest that the Arctic ice cap is now 40 percent thinner than it was in the mid-20th century, and its surface area is also shrinking year by year.

ARCTIC FOX

THE ARCTIC FOX IS SIMILAR in appearance to the red fox, *Vulpes vulpes*, except that it is a little smaller. Its ears and muzzle are also shorter, so that it looks almost catlike. In winter its coat becomes very long, both this and the short ears being adaptations to living in the far north. The small surface area of the ears prevents excessive loss of body heat, while the thick coat acts as an excellent insulation, keeping heat in.

In summer the Arctic fox's coat is grayish yellow, with white on the underparts. In some forms the entire pelt turns white or cream in winter. The blue fox is a variety of Arctic fox that has a bluish gray coat throughout the year. The proportion of blue foxes varies in different regions. They are common around coasts and on islands, where there is less snow in the winter.

The Arctic fox has adapted further to life in polar regions in that long hairs grow on the soles of its feet, a characteristic shared with the polar bear. The hairs help the foxes to keep their footing on ice, as well as providing insulation.

From Arctic Circle to the Pole

The range of the Arctic fox covers the treeless tundra that extends round the Arctic regions of Europe, Asia and North America, and includes Greenland, Iceland and Scandinavia, where foxes are found in the mountains around the northern coasts of Finland and Norway. In winter they move farther south, reaching the province of Quebec, in Canada, and southern Norway and Sweden, in Europe. Arctic foxes can also be found on the smallest and most remote islands north of Canada and Greenland, where there are no other land mammals except polar bears. The foxes reach these almost inaccessible places by traveling across the pack ice, swimming between the ice floes when necessary. They have been found on pack ice within 300 miles (480 km) of the North Pole, where they apparently feed on the remains of seals killed by polar bears and on fish.

Opportunist feeders

Arctic foxes live a communal and nomadic life, often forming small bands ranging the countryside in search of food. They are often less wary than the red fox and in remote areas show no fear of humans. They do not hibernate and are able to withstand very low temperatures.

Arctic foxes eat a wide variety of food, depending on where they live and the time of year. In the European sector the main food of Arctic foxes is lemmings. Where lemmings and

Many forms of Arctic foxes turn completely white in winter. Others, living where there is less snow, may be a bluish gray color throughout the year.

An Arctic fox pup, St. George Island, Alaska. When prey, particularly lemmings, is in great abundance, as many as 20 or 25 young foxes may be born in a litter.

ARCTIC FOX

CLASS	**Mammalia**
ORDER	**Carnivora**
FAMILY	**Canidae**
GENUS AND SPECIES	*Alopex lagopus*

WEIGHT
5–13 lb. (2.3–6 kg)

LENGTH
Head and body: 1⅔–2½ ft. (50–75 cm); tail: 10–16 in. (25–40 cm)

DISTINCTIVE FEATURES
Smaller than red fox with shorter muzzle; some forms white in winter, others bluish gray throughout year; thick, bushy tail

DIET
Small mammals such as lemmings and voles, carrion, birds and eggs; also insects, fish mollusks, fruits and berries

BREEDING
Age at first breeding: 10 months; breeding season: March–April; gestation period: about 52 days; number of young: 2 to 25, depending on food availability, but usually 5 to 8; breeding interval: 1 year

LIFE SPAN
Up to 11 years

HABITAT
Arctic and alpine tundra; also coastal areas, pack ice and areas near human habitation

DISTRIBUTION
Circumpolar range in Arctic: northern regions of Europe, Asia and North America, including Greenland, Iceland and Scandinavia

STATUS
Generally common across most of range; rare in Scandinavia

voles, another major item of food, are scarce they will feed on hares, fish, fruits and berries. Sometime they roam the seashores, feeding on shellfish and carrion. Arctic foxes are also the main predator of many birds, especially ground-nesting species such as ducks, gulls and shorebirds, mainly taking eggs and young.

During the summer, when food is abundant, the Arctic foxes kill more than they immediately need. The surplus is carried back to their dens where it is stored for use during the winter.

In the pack ice of the far north, the Arctic fox follows the polar bear to feed on the leftovers from kills. In the spring the polar bears' seal hunting is thought to be essential to the survival of the foxes, as there is no other food available.

Adjustable reproduction

The breeding season begins in April and the cubs are born in May or June after a gestation of 52 days. The usual litter is of five to eight pups, but when there is a great abundance of food, litters of 20 or 25 young have been known. The male parent stays with the family, helping to feed the cubs.

Friend or foe?

Polar bears sometimes attack Arctic foxes, especially if they are very hungry, and they may well lash out with a paw if the foxes come too close while they are feeding. The foxes may also attack and kill one of their own number under the same circumstances, or if one is injured.

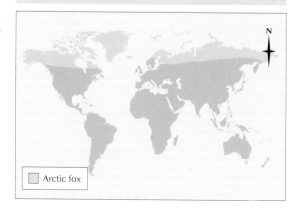

Arctic fox

ARMADILLO

THERE ARE 20 SPECIES of armadillos, grouped in eight genera. They all have a protective covering of armor, brown or pinkish in color, and are stout, short-legged animals with strong claws. Armadillos are distributed through the tropical and subtropical Americas from Argentina to the southeastern corner of the United States. Most live in open areas but some are found in forests. Although they belong to the order Edentata, meaning "no teeth," armadillos do have primitive teeth.

The three-, six- and nine-banded armadillos belong to the genera *Tolypeutes*, *Euphractus* and *Dasypus*, respectively. They are named for the number of movable bands on their armor. The best known of these is one species of nine-banded armadillo, *Dasypus novemcinctus*. It ranges northward from South America into Kansas and Missouri in the United States.

Armadillos, large and small

The giant armadillo, *Priodontes giganteus*, of the forests of eastern South America, is the largest species. It has a 3-foot (90-cm) body and can weigh up to 130 pounds (60 kg). It is unusual in having up to a hundred small teeth, more than twice the normal number for a mammal.

This compares to the smallest armadillo, the lesser fairy armadillo or pichiciago, *Chlamyphorus truncatus*. It is just 9 inches (23 cm) in length, including its 3-inch (7.5-cm) tail, and weighs 4 ounces (113 g). This species, found only on the plains of western Argentina, has less armor than the others. The carapace (back armor) is made up of bands hinged together and is attached to the armadillo's body only by a narrow ridge of flesh running down its spine. There is another flat shield consisting of a single plate covering its rump, and its armored tail sticks through this. The rest of its body is covered with a fine, soft, white fur. The fairy armadillo is molelike, having powerful front legs and small eyes. It spends more time underground than other armadillos.

Roll into a ball

The naked-tailed armadillos, genus *Cabassous*, of Central and South America, have five large claws on their front feet. The middle claw is especially large and sickle-shaped. The three-banded armadillo or apara, *Tolypeutes matacus*, of Bolivia, Argentina and Brazil, is the only armadillo to have its carapace separated from the skin around the sides of its body. As a result it is also the only species that is able to roll up into a complete sphere. The separation of the carapace from the

skin means there is room for its head, legs and tail when it rolls up. Other species are able to roll up to some extent, but not so completely.

The pygmy armadillo or pichi, *Zaedyus pichiy*, is a common resident of Patagonia and the Argentinian pampas. It is said to hibernate.

Dig burrows

Armadillos are mainly nocturnal and live in burrows when not active. They might be found alone, in pairs or in small groups. Nine-banded armadillos will only share burrows with other animals of the same sex. Armadillos are good at digging and their burrows are usually 2–3 feet (60–90 cm) beneath the surface.

Rivers are no obstacles to armadillos, for although they are proportionately heavy due to their coats of armor, they gain added buoyancy by swallowing air to blow up the intestine. The nine-banded armadillo, for example, is said to be able to submerge for 6 minutes.

Some species of armadillos have an unusual gait. The soles of their hind feet are pressed to the ground as they walk, but their forefeet are raised up on the strong pointed claws.

The nine-banded armadillo, Dasypus novemcinctus, *is the only armadillo found in the United States. Its behavior and habits are better known than those of other species.*

The hairy armadillo uses its head as a drill to get at grubs and other insects, forcing it into the ground, then twisting its body to make a hole.

ARMADILLOS

CLASS   **Mammalia**

ORDER   **Edentata (alternatively Xenarthra)**

FAMILY   **Dasypodidae**

GENUS   **Nine-banded armadillos, *Dasypus*; six-banded armadillos, *Euphractus*; three-banded armadillos, *Tolypeutes*; fairy armadillos, *Chlamyphorus*; naked-tailed armadillos, *Cabassous*; hairy armadillos, *Chaetophractus*; giant armadillo, *Priodontes*; pygmy armadillo, *Zaedyus***

SPECIES   **20 species**

WEIGHT

Fairy armadillos: 3–4 oz. (85–113 g). Giant armadillo: up to 130 lb. (60 kg).

LENGTH

Lesser fairy armadillo. Head and body: up to 6 in. (15 cm); tail: 3 in. (7.5 cm). Giant armadillo. Head and body: up to 3 ft. (90 cm); tail: 1⅔ ft. (50 cm).

DISTINCTIVE FEATURES

Armor-plated back and head; soft fur on underparts; protruding ears; short legs with powerful claws; tapering, conical tail

DIET

Invertebrates, lizards, snakes, rodents, carrion, fruits and other plant matter

BREEDING

Gestation period: delayed implantation, then up to 120 days, according to species; number of young: 1 to 4 (most species)

LIFE SPAN

Varies according to species

HABITAT

Savanna, pampas (grassland) and forest

DISTRIBUTION

Southern U.S. to southern South America

STATUS

Some species common; 6 species threatened

Omnivorous feeders

Armadillos live on a wide variety of food, such as insects and other invertebrates, plants, carrion and small vertebrates such as snakes and lizards. Naked-tailed armadillos feed mainly on ants and termites, cutting open their nests with their sickle-like claws and extracting the insects with their long, extensible tongues.

Legend has it that the giant armadillo sometimes digs into new graves to get at human corpses. The peludo or hairy armadillo, *Chaetophractus villosus*, is known to burrow under, and sometimes into, carcasses to get at maggots. It will also dig into soft soil for grubs and other insects in a most unusual manner. It forces its head into the ground, then twists its body round to make a conical hole. These armadillos have also been seen killing snakes by cutting them with the hard edges of their carapaces.

Delayed implantation

Except for the nine-banded armadillos, breeding habits are not well known. Male armadillos mark their home range with urine, in much the same way as a domestic dog or cat. After mating, females exhibit delayed implantation, that is, the development of the embryo does not take place immediately. Instead, the single egg is fertilized and then lies free in the uterus for a period of time before becoming embedded in the uterine wall. At this point development can continue.

Gestation periods vary with species thereafter. In nine-banded armadillos mating takes place in July and August, the female lying on her back during courtship. Gestation takes around 120 days in this case, but is known to be just 65 days in the hairy armadillo.

Armadillos

Identical young

One to four young are normally born each year, depending on the species. In some species the female might bear up to 12 identical young, all of which develop from a single egg.

In the nine-banded armadillos there are four in a litter and, as in other armadillos, the young are all identical, in sex as well as other characteristics. Armadillo young of the same litter are always identical because they all come from the same fertilized egg. All are attached by umbilical cords to a single placenta. This is the area of the uterine wall specialized for transferring nutrients between the blood of the mother and that of the embryos. In other mammals such multiple births are accidental and therefore rare, but identical young are the rule in armadillos.

The young are born with a soft leathery skin that hardens with age. They reach sexual maturity at around 1 year of age in the three-banded species. Life spans vary with species. Giant and nine-banded armadillos live for some 12 to 15 years. One six-banded armadillo lived for 18 years in captivity.

Protective armor

The name armadillo is derived from the diminutive of the Spanish word *armado*, meaning one that is armed. Body armor in mammals is generally made of compressed hair, as in the plates of pangolins and the horns of rhinos, but the armor of armadillos is made up of small plates of bone, each covered by a layer of horny skin and separated from its neighbors by soft skin, from which sparse hairs grow.

The carapace, or back armor, hangs down over the body, protecting the soft underparts and limbs. It is divided into two shields, one covering the forelimbs and one the hind limbs, the two being linked across the middle of the back by a series of transverse bands of plates that allow the carapace to be flexed. The number of transverse bands varies between species. In some they are sufficiently flexible to allow the animal to curl up. The head is also armored and in most species the tail is protected by a series of transverse, bony rings. The softer underparts are covered with a dense layer of hair and scattered, small bony scales.

Defensive behavior

If cornered, armadillos will defend themselves with their sharp claws, but they are more likely to run away, some species moving surprisingly

fast. They will also attempt to burrow into the ground if they cannot find a hole. Armadillos such as the pichi will draw in their feet and wedge the surrounding carapace firmly into the ground. This ruse is effective against birds and some mammals, but not against coyotes, which can pierce their armor. The three-banded armadillo is more effectively protected by being able to roll itself into a complete sphere. Nonetheless, large predators such as the jaguar have a large enough gape to crack even this protective shell.

Becoming rare

Some species of armadillos are agricultural pests, tearing up crops in search of insects, but they can also be beneficial to farmers because they eat unwanted insects. This is not their only use to humans. Armadillos are the only other known mammals to carry leprosy, so are often used in medical research and vaccine development.

In addition, armadillos are hunted across their range for their flesh. Their armor has also been used to make items such as baskets, and the nine-banded armadillo, in particular, is increasingly meeting the hazard of motor traffic.

These factors and others are starting to have an impact on armadillo populations. The already rare lesser fairy armadillo is becoming rarer with the spread of agriculture. The giant armadillo, greater fairy armadillo, *Chlamyphorus retusus*, and several other species are also listed as being vulnerable or endangered as a result of disturbance and persecution by humans.

Reproduction in armadillos is unusual in that the young of a litter all spring from a single fertilized egg. As a result the offspring are identical in sex and other characteristics.

ARMORED CATFISH

THERE ARE MORE THAN 650 species of armored catfish, all of which live in the streams of South America. They are small, the largest being less than 1 foot (30 cm) long, and they are remarkable for two things. One is their bony armor, the other is the method of fertilization used by most species.

Armored catfish belong to three separate families, the first of which, with about 90 species, is known as the thorny catfish. The second, the mailed catfish, contains 130 species with a smooth armor of two rows of overlapping bony plates on each side of the body. The third family, the loricariid catfish, includes about 550 species in which the whole body is covered with overlapping scales. All have barbels, the "whiskers" that give catfish their name.

Talking habits
Mailed catfish can travel overland for considerable distances, pulling themselves along with the strong spines on the breast or pectoral fins and using intestinal respiration, like that in the thorny catfish. Both have a supplementary breathing system, in which air is swallowed and the oxygen from it taken up by a network of fine capillary blood vessels in the wall of the intestine. One species of thorny catfish is known as the talking catfish because both in and out of water it may make a grunting sound. This is caused by movements of the spines in the pectoral fins, amplified by the gas-filled swimbladder acting as a resonator. Several other catfish also make these sounds.

Predatory lifestyle
Mailed catfish live in small groups in slow-moving streams, rarely in standing water, and feed on small animals, such as water fleas. Thorny catfish are active mainly at twilight, when they grub on the bottom for insect larvae and worms. Both feed on small pieces of carrion.

The loricariid catfish are bottom dwellers, mainly in mountain streams or swiftly flowing lowland streams. They have thick lips that form suckerlike mouths, with which they cling to stones and water plants. This serves two purposes: to maintain their position against strong currents and, while doing so, to feed by scraping small algae from the surfaces of the stones and plants with their bilobed or spoon-shaped teeth.

Spawning with the mouth
Little is known about reproduction in the thorny catfish, and although they have been kept in captivity, none has been seen breeding. One of the mailed catfish in the genus *Hassar* will spawn in captivity only when water is sprayed onto the surface of the water in the aquarium, simulating a tropical shower.

There is some disagreement about which of two mating methods is used by the mailed catfish. Some observers say that the male grips the barbels of the female with his strong pectoral fins, so that the two lie with their undersides opposed. In this position the male pours out his milt (sperm-containing fluid) as the eggs are extruded, thereby fertilizing them.

The second method that has been described is one in which the female takes the milt directly from the male into her mouth. There is a certain amount of ritual courtship beforehand in which the male nudges the female with his snout and the two swim over to the surface of a stone, which they clean by removing the minute growths of weed and debris with their mouths. The nudging and cleaning alternate so that by the time pairing takes place they are in a highly excited state and there are several clean surfaces. The purpose of the female sucking in the milt now becomes clear. She swims over to one of the cleaned surfaces. The milt then streams out across her gills as she breathes and the current of water carries it over the eggs as she deposits them on the clean surface, where they adhere.

Corydoras armatus, a member of the Callichthyidae family or mailed catfish, has protective overlapping scales that run along the sides of the body.

ASP

THE ASP, ASPIC VIPER or June viper is closely related to the European adder, *Vipera berus*, and is a member of the family Viperidae. Adders and asps are similar in color, although the ground color of the asp is lighter, usually gray, gray brown, coppery red or orange. Its underparts are gray, dirty yellow or blackish, with a sulphur-yellow or orange-red patch under the tip of its tail. The upper part of the asp's body is often marked with transverse, dark brown or black bars, sometimes zigzags, and occasionally there is an inverted "V" on the head. Differences in size and color between individuals are very much less marked than in the adder.

The neck of the asp is more slender than that of the adder and the species rarely attains a total length of more than 2 feet (60 cm). The largest recorded individual was 2⅕ feet (67 cm) long.

Upturned snout

The adder and the asp can be distinguished by looking closely at the head. The shields on the asp's head are small, and the iris of its eye is shiny yellow, as compared with the coppery red of the adder. Furthermore, the tip of the asp's snout is turned up to make a small spike. This feature is more conspicuous in two related species: the long-nosed viper, *Vipera ammodytes*, of southern Europe, and Lataste's viper, *V. lastastei*, of Spain and northwest Africa.

The asp is common in many parts of Europe, generally farther south than the adder, but where the two ranges overlap it is often difficult to decide to which species a specimen belongs. Hilly or mountainous country is especially favored by asps and the snakes are often found at high altitudes. They have been recorded at 9,700 feet (2,955 m) in the Alps.

The habitat of the asp is generally warm and dry. It frequents rocks, wasteland, hedges and scrub. Each individual has a small home range of several square yards, which it rarely leaves. Asps are active both by day and by night, retiring at irregular intervals to a hole in the earth or between rocks. In winter they hibernate, sometimes several individuals coiling together in one shelter, or hibernaculum.

Aggressive snakes

The asp is a slow-moving snake, but it can be aggressive and dangerous to humans. Incidents are fairly common and bites have proved fatal. Nevertheless, even in the south of France, where the asp is common and there are probably more cases of snakebite than anywhere else in Europe, venomous snakes are not such a danger to people as they are in tropical regions.

Asps feed mainly on small mammals such as mice and voles, young birds and lizards. The very young snakes eat earthworms and insects.

Asp basking on a wall, France. Asps, which can be aggressive and often bite when provoked or alarmed, are specially common in the south of France.

The asp is generally lighter in color than the similar adder. In addition its snout is upturned into a small spike and it ranges farther south.

Ritual breeding battles

Breeding is variable, depending to an extent on climate and location. Pairing of the male and female snakes usually takes place between April and May. Prior to mating, the males indulge in ritualized battles, while the females look on. First, the males attempt to intimidate each other by rearing up in an "S" shape, then, if neither retreats, they chase each other and try to coil round each other's bodies. They never attempt to bite and neither is ever harmed.

Asps are ovoviviparous, the eggs being retained in the mother's body until they are due to hatch. Sometimes the egg membrane ruptures while still in the oviduct and the young are born alive. After a gestation period of around 120 to 150 days, the female produces between 4 and 18 young in August or September. The young measure 7–8 inches (18–20 cm) at birth.

Cleopatra's asp

The asp is best known for being the species of snake with which Cleopatra is believed to have killed herself. Yet it is hardly likely that she would have used the asp, *Vipera aspis*, as this species does not live in Egypt. The reason for Cleopatra's snake being called an asp is that in past times the name was given to any kind of venomous snake, in much the same way that "serpent" can used to describe any snake.

The drawback to a member of the viper family being employed for suicide is that, even when deliberately encouraged, their bites are not often fatal. What is more, the effects of viper venom are usually painful and messy. The venom of vipers is a systemic poison, which clots the blood and destroys the lining of the blood vesssels. The venom of cobras, on the other hand, is a fast-acting poison that interferes with the action of nerves and muscles. Herpetologists, specialists in the study of reptiles, have argued that Cleopatra's asp was most likely to have been the Egyptian cobra, *Naja haje*.

ASP

CLASS	**Reptilia**
ORDER	**Squamata**
SUBORDER	**Serpentes**
FAMILY	**Viperidae**
GENUS AND SPECIES	***Vipera aspis***

ALTERNATIVE NAMES
Aspic viper; June viper

LENGTH
Head and body: up to 1⅔ ft. (50 cm); tail: up to 2½ in. (6.5 cm)

DISTINCTIVE FEATURES
Gray, gray brown, coppery red or orange overall; gray, grayish yellow or blackish underparts; often has dark brown or black markings, bars or zigzags; occasionally has inverted "V" marking on head; snout upturned into small spike

DIET
Adult: small mammals, young birds and lizards. Young: earthworms and insects.

BREEDING
Age at first breeding: 3–4 years; breeding season: normally April–May; gestation period: 120–150 days; number of young: 4 to 18; breeding interval: 1 or 2 years

LIFE SPAN
Probably 10–15 years

HABITAT
Heathland, clearings in deciduous and pine forests, meadow edges, rough grassland, hedgerows and roadside verges; often on hills or mountains

DISTRIBUTION
Southern Europe, including northern Spain, Italy and southern France

STATUS
Common

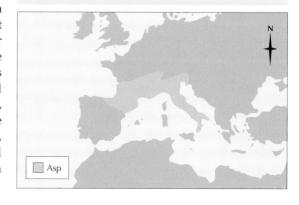

Asp

ATLANTIC SALMON

With a slim, streamlined body, the Atlantic salmon is obviously built for speed, and this is confirmed by its long, leaping and powerful struggle when caught on rod and line. This salmon is a popular fish, fished by humans both for food and sport. Its numbers have been reduced to critically low levels in some regions as a result of this, although pollution and the blockage of rivers and streams by dams have also featured in the Atlantic salmon's decline. Salmon farming has been developed in many areas to raise the fish in captivity, but this strategy is itself creating problems for wild populations.

The Atlantic salmon reaches 3⅓–4 feet (1–1.3 m) in length, exceptionally up to 5 feet (1.5 m). It is troutlike in appearance, its body elongated with slightly compressed sides. The upper and lower rays of the tail stand out from the outline, the tail itself being shallowly forked.

Coloration depends very much on age, whether the fish is at sea or in fresh water, and whether or not it is ready to spawn. Smolts (young salmon about 2 years old) and adult salmon returning from the sea are green or blue on the back, silvery on the sides and white below. As the adult approaches spawning it becomes darker: brown or bronze with red spots and dark fins. Fish that have spawned are very dark in color, often with heavy red patches. The parr (young fish) are dark above with a series of 8 to 11 dark, rounded parr marks on the sides.

Breeds in fresh water

The Atlantic salmon is one of the family Salmonidae, which consists of only five genera, and perhaps two dozen species, including the familiar brown, cutthroat and rainbow trout. The salmon exhibits anadromous (from the Greek for running upwards) behavior, meaning that it feeds in the sea and breeds in fresh water. It is a matter of opinion whether the species was originally marine and took to migrating into rivers to spawn, or was a freshwater fish that took to going down to the sea to feed. Most evidence is in favor of the first explanation.

As the name suggests, these salmon are found on both sides of the Atlantic. In the Western Atlantic they occur within the coastal drainages from northern Quebec in Canada to Connecticut in the United States. In the Eastern Atlantic they are found in the drainages from the Arctic circle to Portugal. There are also landlocked stocks present in Russia, Finland, Sweden and Norway, and in parts of North America.

A thousand-mile migration

Most of our knowledge of the habits of salmon concerns their stay in fresh water. What happens to salmon in the sea is something of a mystery, although it has been a focus of interest for scientists for many years. Results from tagging show that the salmon may travel up to 1,000 miles (1,600 km) or more from the river mouth where they enter the sea, although most of them wander less than 100 miles (160 km) from it.

A salmon returns to spawn in the same river in which it was hatched, and there has been a great deal of speculation and research on how this is done. How the salmon makes its way from its feeding ground to the river where it will spawn is not known for certain. However, once it has found its way to the home river, it reaches the pool or stream where it first started life by using its keen sense of smell.

As soon as the salmon has entered its ancestral river, it has the urge to go upstream no matter what the obstacles. So we have the famous

Mature salmon, such as this one in Scotland, return from the sea to spawn in the same river in which they were hatched. They will leap any obstacles, for example waterfalls, that they encounter along the way.

Salmon alevins, or recently hatched juveniles, 4 days after hatching. At this stage they still carry yolk sacs, which are gradually absorbed.

leaps up waterfalls, 10-foot (3-m) leaps being recorded. Where a river is dammed, as for a hydroelectric station, fish ladders are often built, in the form of a series of steps, to help the salmon to make their way upstream.

Color changes after spawning

When they travel upriver, the salmon are in good condition, their flesh firm and red and the surface of the body silvery. There is much fat stored in the body but later, as a result of spawning, this is used up. At this stage, the flesh becomes pale and watery and the outside of the body loses its silvery appearance, becoming darker. The skin of the back becomes thick and spongy, with the scales deeply embedded. Large black spots margined with white appear on the body, which is spotted and mottled with red and orange. This produces the "red fish," which are males, the females being similarly colored but darker. The males are further distinguished by the way the snout becomes longer and the lower jaw hooked.

Food into red flesh

During their stay in the sea, the salmon spend much of their time at moderate depths, which may account for the infrequency with which they are caught in trawls. They probably move nearer the surface at night, following the plankton on its daily migration upwards, since they feed on various shrimplike crustaceans. It is from the pigments in these, especially the carotenoids, that the salmons' flesh derives its pink color. In addition, sand eels, squid, small herrings and other fish are taken by the adult fish at sea. When in fresh water no regular feeding takes place, the salmon drawing on their reserve food store of fat.

ATLANTIC SALMON

CLASS	**Osteichthyes**
ORDER	**Salmoniformes**
FAMILY	**Salmonidae**
GENUS AND SPECIES	***Salmo salar***

WEIGHT
Up to 100 lb. (45 kg), usually less

LENGTH
Up to 5 ft. (1.5 m), usually less

DISTINCTIVE FEATURES
Elongated, troutlike body; shallowly forked tail. Smolt (2-year-old fish) and adult returning from sea: green or blue back, silvery sides and white belly. Adult at and after spawning: darker with red spots or patches. Parr (young fish): dark above, with 8 to 11 parr marks on sides.

DIET
Adult: squid, shrimps and fish at sea; does not feed in fresh water before breeding. Juvenile: mollusks, crustaceans, insects and some small fish.

BREEDING
Age at first breeding: about 2 years; spawning season: November–December; number of eggs: 800 to 900 for every pound of female's weight; hatching period: 5–21 weeks

LIFE SPAN
Up to 13 years

HABITAT
Spawning grounds: clean streams and rivers, especially with gravel beds. Feeding and breeding grounds: North Atlantic.

DISTRIBUTION
North Atlantic; fresh waters of northeastern North America and northwestern Europe

STATUS
At low risk

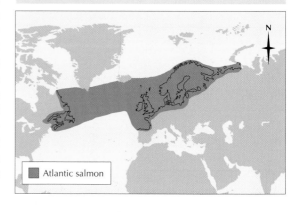

Atlantic salmon

From gravel stream to ocean deep

The Atlantic salmon breeds in the rivers of Europe, from Spain north to the White Sea, off Iceland, in the southern tip of Greenland and in North America, from Labrador in eastern Canada southward to the New England coast.

The life cycle begins in the shallows of a stream where the water is clean and there is a gravel bed. Adults enter estuaries from spring onward, ceasing to feed thereafter, and ascend unpolluted rivers to spawn between November and December. Prior to spawning the female digs depressions in the gravel using lashing movements of her body. The eggs are laid in these troughs, the total number each season being around 800 to 900 for every pound of the female's weight. The eggs are fertilized by the attendant male, shedding his milt (sperm-containing fluid) over them, after which the female covers the eggs with gravel and moves upstream to repeat the process. The eggs are large, about ⅕–⅓ inch (6–8 mm) in diameter, and hatch after between 5 and 21 weeks, depending on temperature.

The recently hatched juveniles, or alevins, still carry a yolk sac and are ½ inch (1.3 cm) long. They remain among the pebbles of the spawning ground until the yolk sac is absorbed, that is, until about 4–8 weeks after hatching. They then leave for shallow water, about 1–2 inches (2.5–5 cm) long at this stage. At this point they are called fingerlings. The juveniles feed on mollusks, crustaceans, insects and fish. After a year, when 3–4 inches (7.5–10 cm) long, the fingerlings become parr, and by the end of the second year they reach a length of 4½–8 inches (11.5–20 cm). At this stage the body is marked with 10 or 11 dark blotches, called parr marks, on each side of the body. There is a single, reddish orange spot between each dark blotch.

The time at which the various stages are reached varies with temperature and with factors such as latitude. The discrepancies become even more pronounced in later stages of the life cycle, for example when the parr becomes more silvery and, as a smolt, is ready to go to sea. For example, in southern England, the smolt stage may be reached in 1 year, while in northern Scandinavia, it may be 7 or even 8 years before the parr becomes a smolt.

On reaching the sea, the salmon feeds for between 1 and 6 years before coming back to the same river to spawn. After spawning, the kelts (spent salmon) drop down river tail-first, weakened by fasting and spawning, often attacked by disease. Only a minority of adults survive the first breeding season; many, especially the males, die on the way back to sea. Those that reach the sea soon recover and start to feed.

Overfishing and pollution

In terms of natural predation, eels take the eggs of Atlantic salmon, and many birds as well as perch, pike and trout feed on the young fish. Otters will take quite large adults. In the sea the main predators are seals, porpoises and cormorants, and some of the larger predatory fish.

However, the main threats to the Atlantic salmon are overfishing by humans, the damming of rivers for irrigation and electricity provision and the pollution of freshwater estuaries and streams by effluents from factories and pesticides washed from agricultural land. In addition, the tagging of salmon has revealed their marine breeding grounds, allowing the fish to be taken by trawlers for the first time. This has depleted their numbers at yet another stage of the life cycle.

Atlantic salmon are now successfully farmed. However, farmed salmon can escape and compete with wild fish for food and spawning grounds. Also farmed salmon, reared from imported eggs, can transmit diseases to wild fish. If they interbreed, wild fish may lose their own genetic adaptations to local conditions. Genetic modification of captive salmon, including crossbreeding with other species, along with the use of growth agents, presents a further threat to wild populations should the captive fish escape.

Atlantic salmon are a popular fish, taken for both food and sport. Overfishing, pollution, the damming of rivers and new fish-farming practices each present a serious threat to wild populations.

ATLAS MOTH

ATLAS MOTH

PHYLUM	**Arthropoda**
CLASS	**Insecta**
ORDER	**Lepidoptera**
FAMILY	**Saturniidae**
GENUS AND SPECIES	***Attacus atlas***

LENGTH
Wingspan: 6¼–12 in. (16–30 cm)

DISTINCTIVE FEATURES
Adult: very large size; curved wingtips; triangular, translucent patches on wings; red flash on curved tip of forewings; pale-edged brown spots along wing margins; brick-red line runs from forewings to hind wings. Larva: rows of spines; covered with white, waxy substance.

DIET
Adult: does not feed. Larva: willow (*Salix*), poplar (*Pupulus*) and privet (*Ligustrum*).

BREEDING
Larval period: 2–3 months; larva pupates in large, tough, silk cocoon

LIFE SPAN
Not known

HABITAT
Lowland tropical forest

DISTRIBUTION
India and Sri Lanka east to China and Philippines, south to Malaysia and Indonesia

STATUS
Uncommon

The atlas moth (male, above) is distinctive for its large size and strikingly patterned, strongly curved forewings.

PROBABLY THE MOST FAMOUS of the large moths, the atlas moth belongs to the family Saturniidae, which includes the British emperor moth and the North American cecropia moth. These species are all known for producing a large silk cocoon. The atlas moth is found in the Tropics and subtropics, from Sri Lanka and India, including the Himalayas, eastwards to China, Malaysia and Indonesia.

The female atlas moth is larger and heavier than the male, and her "hooked" forewings span from 6¼ to 12 inches (16–30 cm). The atlas is the world's largest moth in overall size, although the owlet moth, *Thysania agrippina*, has a greater wingspan. Its body and wings are generally tawny in color. The wings are strikingly patterned, and there is a conspicuous, triangular, translucent patch on each wing.

Habits and life history

The ovoid eggs are laid in clusters on a wide variety of tropical shrubs, including cinnamon and hibiscus. The larvae or caterpillars are at first white but later turn a pale bluish or yellowish green. The caterpillar's body is ornamented with rows of spines and is covered with a white, waxy substance. Two, sometimes three months after the eggs are laid, the caterpillar, which is now 4–4½ inches (10–11.5 cm) long and over 1 inch (2.5 cm) in diameter, spins a tough cocoon of silk around itself, in which it pupates. This cocoon is attached to the foliage of the food plant. When it is ready to emerge, the moth secretes a liquid from the mouthparts, which dissolves the silk and enables the moth to push its way out. The cocoons are very thick and the moth needs it to be moist before it can emerge.

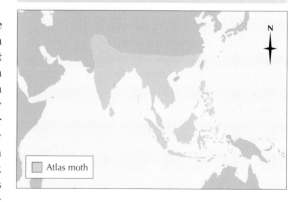

Atlas moth

The atlas moth can be kept in captivity in temperate latitudes, the larvae feeding readily on privet, willow and poplar, but a temperature of 70–80° F (21–26° C) needs to be provided. Adult moths are nonfeeding and lack mouths.

AVADAVAT

THERE ARE TWO species of avadavats: the globally threatened green avadavat, *Amandava formosa*, which is endemic (restricted) to India, and the more widespread red avadavat, *A. amandava*, which is found across much of India and Southeast Asia. The name avadavat is a corruption of Ahmadabad, the Indian city from which the birds were first sent to Europe. The red avadavat is sometimes called the strawberry finch, but the name is not restricted to this species. Red avadavats are about the size of house wrens, *Troglodytes aedon*, but are strikingly colored. The male is coppery to bright red with black underparts, dark brown wings and tail, and a reddish brown crown. The back, rump, wings and belly are dashed with white spots. The female is more somber, being medium brown with paler underparts and buffish on the belly, with some red on the tail and rump and white spots on the wings. The male red avadavat is unique in the finch family for having a nonbreeding "eclipse" plumage: at the end of the breeding season the male molts to a plumage very much like that of the female, although he can always be distinguished from her by the white spotting on the red of his rump. The green avadavat is olive green above with yellow underparts.

In breeding plumage the male red avadavat is bright red with distinctive white spotting on the wings, belly, back and rump.

Clumping together

Avadavats live in damp areas where reeds and tall grasses flourish. When they are not feeding or busy with nesting, avadavats spend their time sitting together in small groups, usually of three or four, and sometimes up to 30. Like the anis, a group of sociable American birds in the genus *Crotophaga*, the avadavats are very tolerant of each other's company and they do more than merely sit together: they actively "clump." That is, they push or lean against each other. Clumps usually start to form around a bird that is sitting quietly. Other birds approach it carefully in a submissive attitude so that it is not provoked into attacking or fleeing.

Diet and plumage

In the wild, avadavats feed on the ground, picking up seeds and insects. They also do well and are popular as caged birds, but it is important that they are fed the right foods or they will lose their brilliant red plumage and become brown or even black. A proper diet for avadavats includes seeds, such as small yellow millet, fresh seeding or flowering grasses and ants. All these foodstuffs contain carotenoids, members of the group of chemicals to which carotene, which gives the orange color to carrots, belongs. The birds convert these chemicals into brighter forms, which are concentrated in their feathers.

Post-monsoon breeding display

In the wild avadavats breed after the monsoons. Before mating they go through a series of displays. One is called the straw display, in which either the male or the female picks up a feather or piece of grass by the shaft and, holding it out in front of itself, fluffs out its feathers and bows slowly. The displaying bird, be it male or female, also sings during the display, a high-pitched warble that slowly descends the scale.

Each male defends an area extending several yards in each direction from his nest. Somewhere in this area will be a special perch where he sings. He shows aggression to any male that comes into his territory, either by displaying at him or by attacking. The signal that releases a cock's aggressiveness is the red plumage on the other cock. This acts in the same way as the red breast of the European robin, *Erithacus rubecula*. Females and cocks in eclipse plumage are ignored. Hens, however, will attack other hens.

The nests of avadavats are untidy. They take bundles of material to the nest site, usually high in a tree. First, they make a flat platform for the nest among the twigs. Then they add grass and finally a lining of feathers. Four to six pure white eggs are laid and incubated for about 13–14 days. The young are fed by both parents. When a young avadavat is approached by one of its parents, it crouches and opens its bill, at the same time waving its head. It also opens both its wings and often flutters them. This performance is known as food-begging and the form it takes depends on how hungry the chick is. If it is well fed the chick opens its bill without begging.

Social preening

Avadavats, and other birds that live together in close groups, can often be seen preening each other. This is called allopreening, to distinguish it from autopreening, when a bird preens its own feathers. In allopreening, the preening bird grasps a feather of the other bird at the base and draws its bill along the shaft, gently nibbling at it as it goes. This is precisely what a bird does when preening its own feathers. Allopreening is, however, usually confined to the partner's head, and the bird being preened assists the preener by rolling and twisting its head to present different parts to the preener's bill.

Allopreening takes place when the avadavats are perched in their tight clumps. Sometimes a bird will lean over its neighbor and preen the bird beyond it. A bird will invite another to preen it by adopting a special "invitation" posture. It ruffles the plumage around its head and rapidly opens and shuts its bill.

A captive pair of red avadavats. This species is a popular cagebird.

RED AVADAVAT

CLASS	**Aves**
ORDER	**Passeriformes**
FAMILY	**Ploceidae**
SUBFAMILY	**Estrildinae**
GENUS AND SPECIES	***Amandava amandava***

ALTERNATIVE NAMES
Avadavat; red munia; strawberry finch; tiger finch; strawberry waxbill

WEIGHT
⅓ oz. (10 g)

LENGTH
Head to tail: about 4 in. (10 cm); wingspan: 5–5½ in. (13–14.5 cm)

DISTINCTIVE FEATURES
Short, broad-based bill. Breeding male: carmine red with white spots. Female and nonbreeding male: brownish overall with some red on tail and rump.

DIET
Grass seeds; occasionally small insects

BREEDING
Age at first breeding: 1 year; breeding season: June–December; number of eggs: usually 4 to 6; incubation period: 13–14 days; fledging period: 17–21 days; breeding interval: not known

LIFE SPAN
Not known

HABITAT
Tall grass and cultivated land; roosts in reed beds and sugarcane beds

DISTRIBUTION
Indian subcontinent and Southeast Asia; introduced to Mauritius, Fiji, Singapore and parts of Spain

STATUS
Fairly common in native range

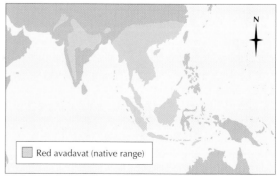

Red avadavat (native range)

AVOCET

SHOREBIRDS THAT BELONG to the family Recurvirostridae, avocets are easily identifiable by their long, upwardly curved bills. There are four species that are widely distributed. The adult American avocet, *Recurvirostra americana*, during the breeding season, is black and white on its upperparts and white below, with a rusty colored head and neck. In the winter it loses this plumage and gray replaces the rusty areas. The Chilean avocet, *R. andina*, and Australian or red-necked avocet, *R. novaehollandiae*, have a similar plumage, but the Chilean species is darker than the American avocet and the Australian species has a more reddish brown plumage. The Eurasian avocet, *R. avosetta*, which is slightly smaller than the American, lacks any reddish or rusty tones at all in its plumage.

The black-and-white plumage of the Eurasian avocet is so strikingly impressive that it has been adopted as a symbol by the Royal Society for the Protection of Birds. Based in Britain, this is the world's largest conservation charity, with more than 1 million members worldwide in 2000.

Life in the salty shallows

The Eurasian, American and Australian avocets breed on mudflats, lagoons and salt marshes around the coast, and sometimes in less saline areas. Outside the breeding season, sandbanks and shoals, or flats around lakes and rivers, are favorite haunts. The Chilean avocet lives on lakes in the High Andes of South America.

In flight, an avocet's neck is only slightly extended so that it looks comparatively short. The legs are held out, stretching well beyond the tail. The wingbeats are rapid and regular until landing, when the bird glides in. After landing the avocet often stands for a short time with its wings extended.

Avocets may be seen singly, but they are gregarious birds, and in winter can be seen in flocks hundreds strong. The call note of the American avocet is a loud *wheet*, while that of the Eurasian avocet is a more musical *klooit*, rapidly repeated if a bird is raising the alarm.

Sweeping for food

The feeding patterns of shorebirds are largely determined by their bill shape. Many species use their long bills to probe the mud and sand along the sea coasts in search of the small animals that live beneath the surface. The curved bill of avocets is an adaptation to a slightly different food source. Avocets live on small crustaceans, worms, fish and mollusks, as well as seeds and other plant material, which float around the shore. An avocet wades through the shallows with its bill held just below the water, the front end parallel to the surface. In this position it moves forward, sweeping the bill from side to side through a 50° arc. A bird with a straight bill would be unable to feed like this. Every now and then the avocet swallows the food that has been caught in its slightly opened bill. In deeper water it holds its head underwater, and it may "up end" like a duck. Avocets often feed in groups, striding forward shoulder to shoulder.

Breeding behavior

The Eurasian avocet breeds on the North and Baltic Sea coasts, around the Mediterranean, Caspian and Black Seas, and in parts of Central Asia and the Middle East. In winter it has a broader range that includes parts of Africa (especially the south) and southern Asia. American avocets breed across much of the United States and southern Canada, moving south after the breeding season to winter on the coasts of California, Texas and Florida, and in Central America.

Avocets nest in small colonies, in which their relatively small nests are sometimes only 2 feet (60 cm) apart. In sandy areas no nest material is used, but when vegetation is available, this is

Avocets (American avocet in summer plumage, below) feed by skimming the water's surface with their long and slender upturned bills.

Avocets become highly excitable during the breeding season, chasing away any other birds that venture too close to their nests. Pictured is the Eurasian avocet.

used in nest-building. The colonies are often subject to sudden flooding, when the avocets rush round collecting material so as to raise the eggs clear of the water. Avocets become highly aggressive when breeding. The breeding pair (it is impossible to tell the difference between the sexes) will attack and chase away any other avocet that dares to intrude.

Mating is accompanied by an elaborate ceremony and nearly always takes place in the water. The female crouches low in the water, while the male walks backward and forward behind her, preening himself continuously. Gradually he draws closer until he brushes against the female's tail. After mating the male jumps to one side of the female, opens his wings and the pair run together, with one of the male's wings held over the female's back.

Between two and eight eggs, usually three or four, are laid in late April, May or June. Both parents share in the 22–25 day incubation period. The chicks leave the nest soon after hatching and fend for themselves. They can catch food when a few hours old and are independent at 10 weeks.

Predators and defense

Avocets are extremely vulnerable to rats and foxes, and to other animals that can take advantage of ground-nesting birds. However, it is the widepread draining of coastal marshes that has led to declines in the numbers of the Eurasian and American avocets in many areas.

In common with many other shorebirds, avocets often show little fear of humans, refusing to move far from their nests when flushed. More likely, they fly out toward a human and swoop over his or her head in an effort to frighten the intruder away from the nest.

EURASIAN AVOCET

CLASS	**Aves**
ORDER	**Charadriiformes**
FAMILY	**Recurvirostridae**
GENUS AND SPECIES	***Recurvirostra avosetta***

ALTERNATIVE NAMES
Pied avocet; black-capped avocet

WEIGHT
9–10¼ oz. (260–290 g)

LENGTH
Head to tail: 16½–17¾ in. (42–45 cm); wingspan: 30–31½ in. (77–80 cm)

DISTINCTIVE FEATURES
Long, slender, upturned bill; very long, bluish legs; bold black-and-white plumage

DIET
Mainly invertebrates, especially insects, crustaceans, mollusks and worms; also small fish and plant matter such as seeds

BREEDING
Age at first breeding: 2 years; breeding season: eggs laid April–June; number of eggs: usually 3 or 4; incubation period: 22–25 days; fledging period: 35–42 days; breeding interval: 1 year

LIFE SPAN
Up to 24 years, usually much less

HABITAT
Breeding: mudflats, saltwater lagoons and salt marshes, mainly on coasts. Nonbreeding: also on sandbanks or flats beside rivers and lakes.

DISTRIBUTION
Coasts and inland saline waters in northern Europe, Mediterranean, Africa, the Middle East and Central and southern Asia

STATUS
Locally common

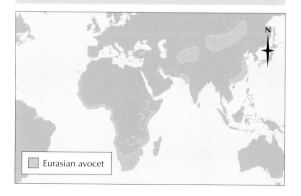

☐ Eurasian avocet

AXIS DEER

THE MOST COMMON DEER in India is one of the four species of the genus *Axis*. It is usually known as the axis deer or by its Hindi name, *chital*. Other alternative names include spotted deer, Ceylon deer and Ganges stag. An attractive deer, the axis deer, *Axis axis*, has a bright, reddish brown coat with lines of conspicuous white spots. Its underparts are white, as are the insides of its ears. Its antlers are slender and have few branches. It is unusual in that the stags (males) shed their antlers at all times of the year, and there tends not to be any particular season for the fawns to be born. The body size of the axis deer varies from one area of its range to another. In northern and central India the stag stands up to 3 feet (90 cm) at the shoulder and weighs up to 178 pounds (81 kg), while in southern India and Sri Lanka it seldom exceeds 2⅗ feet (80 cm) and 165 pounds (75 kg).

Related species

Another species of the same genus lives alone, or in parties of up to 18, on the grassy plains of northern India and Myanmar (Burma), and in some parts of Sri Lanka. It is known as the hog deer, *Axis porcinus*, for its squat, almost piglike appearance. Its legs are short and its body smaller and stouter than that of the axis deer, a stag standing perhaps 2½ feet (75 cm) at the shoulder. It runs with its head down, rather than bounding. Its coat is brown with a yellowish or reddish tinge, and looks speckled because some of the hairs have white tips. Its antlers are small and set on long, bony stalks called pedicels. Despite appearing so different, the hog deer and the axis deer are close relatives. Young hog deer are white-spotted and the two species readily interbreed.

In addition, there are two other species: *Axis calamianensis* and *A. kuhlii*. The first is found only on the Calamian Islands, west of the Philippines, the second in secondary forest on Bawean Island between Borneo and Java. Both are similar in size and appearance to the hog deer.

Lives in large herds

There is no segregation of the sexes except that stags leave the herd when their antlers are about to be shed. Otherwise the axis deer lives in herds of up to several hundred, including stags, hinds and young of varying ages. It is found on lowland

The axis deer or chital *is the most common deer in India and is also found throughout Sri Lanka. It has been introduced, often as a park deer, to many other countries.*

Young axis deer suckling and being washed by its mother. The species is unusual in that fawns are born at any time of year.

plains or the lower hills, among bushes or trees or in bamboo forests, especially near a stream, with a ravine for shelter. The axis deer swims well and readily takes to water. Less nocturnal than most deer, the axis deer feeds for a few hours after sunrise, then goes to water. It rests in shade in the midday heat and feeds again for a couple of hours before sunset. It is both a grazer (grass-eater) and a browser (eater of leaves), and feeds on a wide variety of grasses, trees and herbs.

Fluctuating populations

Mating often takes place from April to May in India, but fawns can be seen at all times of the year, born after a gestation of around 225 days.

As a result of hunting, axis deer have been severely reduced in numbers or even wiped out in some parts of their range. In other areas their numbers have increased because their natural predators are also hunted by humans. Predators are of two kinds. There are those, such as pythons, that take a small but steady toll of the herds, and those, such as the wild dog or dhole and the leopard, which are thought to not only prevent excessive increases in numbers but also keep the herds on the move. This prevents the habitat of the axis deer from becoming overgrazed.

In countries where the deer have been introduced as park deer, and have then gone feral (wild), there may be no natural predators. In this case, there might be a need for culling (the organized killing of a certain number) to prevent overgrazing. Nonetheless, other factors such as the increased incidence of disease, decreased body condition and decreased fertility will tend to keep deer numbers in check should the population be too dense in any one particular area.

AXIS DEER

CLASS	**Mammalia**
ORDER	**Artiodactyla**
FAMILY	**Cervidae**
GENUS AND SPECIES	***Axis axis***

ALTERNATIVE NAMES
Chital; cheetal; cheetul; Ceylon deer; Ganges stag; spotted deer

WEIGHT
Male (in India): up to 178 lb. (81 kg); male (in Sri Lanka): up to 154–165 lb. (70–75 kg)

LENGTH
Shoulder height (male): up to 3 ft. (90 cm); larger in northern part of range

DISTINCTIVE FEATURES
Reddish brown coat with white spots; white patch on throat; dark stripe on back, from neck to tail; slender antlers (male only) curve backward and outward, with few branches

DIET
Grasses, herbs and foliage of trees

BREEDING
Age at first breeding: not known; breeding season: all year, but mainly April–May; gestation period: approximately 225 days; number of young: normally 2; breeding interval: usually about 1 year

LIFE SPAN
Not known

HABITAT
Forest and woodland near permanent water

DISTRIBUTION
Throughout India south of 25° N; Sri Lanka; introduced to Australia, South Africa, Hawaii, South America and North America

STATUS
Locally common in native range

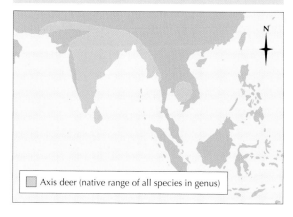

Axis deer (native range of all species in genus)

AXOLOTL

T HE AXOLOTL IS A SALAMANDER, unusual in that it permanently retains larval features. It is also able to reproduce while still in the aquatic larval stage. This is unlike the normal development of amphibians such as frogs, toads and newts, which as larvae, or tadpoles, are confined to fresh water. In the adult form they can live both in water and on land, reproducing in water during the breeding season. Certain amphibians, the Mexican axolotl being the most famous, are able to complete their life cycle without ever leaving the water, as sexual maturity is reached in the larval stage. Such amphibians are said to be neotenic.

The axolotl is a newtlike creature, 4–10 inches (10–25 cm) long. It is usually black or dark brown in color, with black spots, but albinos are also fairly common. Its legs and feet are small and weak, while its tail is long, with a fin running from the back of the head to the tail. There is also a fin along the underside of the tail. The axolotl breathes through the three pairs of feathery, external gills on the sides of its head.

Unusual salamanders

The first known axolotl, the Mexican axolotl, *Ambystoma mexicanum*, was discovered in certain lakes on the central Mexican plateau. The name axolotl is Aztec in origin, derived from "atl" (meaning water) and "xototl" (meaning dog). Later, another species was discovered, *A. dumerilii*, which occurs in the lakes southeast of Mexico City, where it is eaten locally and is considered a delicacy. These two species are completely neotenic.

Around 1865 it was discovered that axolotls belong to the salamanders, a group of amphibians the larval stage of which are aquatic tadpoles resembling adult axolotls. In other salamanders the tadpole later changes into an adult. Axolotls, on the other hand, become sexually mature while still larvae and fail to metamorphose.

The name axolotl is more generally used for any full-grown larva of the genus *Ambystoma* that has not lost its external gills. The other four species of neotenic salamanders are facultative neotenics, meaning that they are neotenics under certain conditions. Of these, the best known are the tiger salamander, *A. tigrinum,* and *A. gracile.* In these cases populations have been known to metamorphose if their aquatic environment becomes inhospitable. This happens as a result of certain hormone changes.

Mating behavior

In most frogs and toads, fertilization of the eggs takes place externally. The female sheds the eggs into the water and the male simply releases his sperm near them, to make their own way to the eggs. The axolotl, related salamanders and newts have a system of internal fertilization. This is different from the normal method, in which the male introduces the sperm into the female's body to meet the eggs waiting there. Instead, the male axolotl sheds his sperm in a packet called a spermatophore, up to 25 such packets being released each night. The spermatophore sinks to the bottom and the female settles over it. The spermatophore is enclosed in a cone of mucus and therefore sticks to the female's body. It enters through the cloaca (a chamber into which the urinary, intestinal and generative canals feed), where it is stored in a receptacle.

Breeding can take place from December, but is normally between March and June. The male attracts the female by a courtship dance, secreting a chemical from glands in his abdomen and swishing his tail. This spreads the chemical

Axolotls (albino Mexican axolotl, below) are in fact salamanders that permanently retain larval features such as external gills.

Completely neotenic axolotls, that is, those that always retain juvenile characteristics in their adult form, are restricted to certain lakes in Mexico.

until a female detects it and swims toward him. Between 24 hours and 1 week later, 200 to 1,100 eggs are laid. They are sticky and the female attaches them to plants with her back legs. The young axolotls hatch out 14 to 21 days later, depending on the temperature of the water. At this stage they are only about ½ inch (1.5 cm) long and remain on the plant where the eggs were originally laid. After a week the young start swimming in search of food and, if the water is warm and food plentiful, they will be 5–7 inches (13–18 cm) in length by winter. The young will then hibernate, taking no food, if the water temperature drops below 50° F (10° C).

Diet

The youngest axolotls feed on plankton, minute organisms that float in water. Later they eat water fleas such as daphnia, and when fully grown they hunt small aquatic prey such as various worms, tadpoles, insect larvae, crustaceans and wounded fish.

Precocious amphibians

The basic cause of neoteny, the retention of juvenile characteristics in the adult form, seems to be a lack of thyroxine, the hormone secreted by the thyroid gland, which controls metabolism. Axolotls in laboratories have been known to change into a gill-less form resembling the adult tiger salamander if given thyroxine. It would seem, then, that there is something lacking in the diet of axolotls. In Wyoming and the Rocky Mountain area of the United States, for example, the tiger salamander regularly exhibits neoteny and humans are liable to get goiters, a swelling of the thyroid gland resulting from low levels of thyroxine. This has been traced to a lack of iodine in the water in these regions, for iodine is an essential component of thyroxine.

MEXICAN AXOLOTL

CLASS **Amphibia**

ORDER **Caudata**

FAMILY **Ambystomatidae**

GENUS AND SPECIES *Ambystoma mexicanum*

ALTERNATIVE NAME
Axototl

WEIGHT
3½–7 oz. (100–200 g)

LENGTH
4–10 in. (10–25 cm), average: 8 in. (20 cm)

DISTINCTIVE FEATURES
Newtlike animal resembling the aquatic larva (tadpole) of a salamander; 3 pairs of feathery external gills; small, weak legs; long tail; fin runs from head to tail; black or brown in color, with black spots; albinos fairly common

DIET
Small aquatic animals such as worms, insects, crustaceans and small fish

BREEDING
Able to reproduce while still in aquatic larval stage. Age at first breeding: usually 2 years; breeding season: from December, but usually March–June; number of eggs: 200 to 1,100; hatching period: 14–21 days; breeding interval: 1 year.

LIFE SPAN
Up to 25 years

HABITAT
Lakes

DISTRIBUTION
Confined to central Mexican plateau

STATUS
Not known

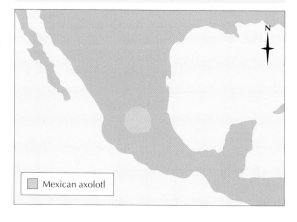

Mexican axolotl

AYE-AYE

OF ALL THE UNUSUAL, and unfortunately rare, animals that live in the forests of Madagascar, the aye-aye is one of the rarest. Many of these animals are lemurs, close relatives of the aye-aye. Originally thought to be a rodent, it is in fact a primate, the sole remaining representative of the family Daubentoniidae.

About size of a domestic cat, the aye-aye is the largest nocturnal primate. It is about 16 inches (40 cm) in length, with a 22–24-inch (55–60-cm), bushy tail. It has a coat of thick, shaggy dark gray or black fur and its face is rounded, with large eyes and erect, naked ears, something like those of a mouse. The fur around its face and throat is yellowish white. The aye-aye could be described as almost squirrel-like, particularly in its teeth, the incisors growing throughout its life as they do in rodents. Despite this, its hands and feet are like those of other primates such as lemurs, monkeys and apes, and have opposable thumbs. Those of the hind feet have flat nails but all the other toes have pointed claws. The middle (index) finger of both its hands is extremely long and narrow. It uses it for feeding, for combing its fur, for scratching and so on.

Nocturnal habits

Aye-ayes live only in northwestern and north-eastern Madagascar, where they inhabit forests, mangroves and bamboo thickets. Occasionally they are found in cultivated areas such as coconut plantations, but it is largely the clearing of land for agriculture that has caused the aye-aye to become so rare. Nonetheless, where they have adapted themselves to life near human settlements they appear to be little disturbed by the general activity.

The aye-aye rarely comes into contact with humans as it is nocturnal. It is also arboreal, spending the day in hollow trees, and foraging at night among the branches. Aye-ayes are mainly solitary and silent creatures, but they occasionally emit a short cry that sounds like pieces of metal being rubbed together.

The Malagasy, the people of Madagascar, regard the aye-aye with dread. In legend, the mere touch of this animal is supposed to cause death. Another legend credits aye-ayes with being the reincarnated ancestors of the Malagasy.

Listening for food

Fruits, wood-boring insect larvae, bamboo shoots and tree sap make up the aye-aye's diet. The two most peculiar features of its anatomy, the elongated middle finger and the rodent- or squirrel-like front teeth, are used for getting food. After nightfall the aye-aye creeps around the branches, listening very carefully for the sound of an insect grub chewing its way through the wood. If it cannot hear anything, it delicately taps the branch with its long index finger, listening, with all the skill of a piano tuner, for any change in sound. Such a change might indicate a hollow where, perhaps, a grub is lurking. If an opening to such a hollow can be found, the aye-aye will insert its long finger and try to hook the grub with its claw, hauling it out of the branch. Failing this, the chisel-like front teeth are used to gnaw away at the wood until the prey can be reached. The aye-aye also uses its elongated index finger to dig the pulp out of fruit.

Aye-aye with young, chewing into a coconut. Little known, the aye-aye is probably one of the rarest and most endangered primates in the world.

Aye-ayes listen for the sound of wood-boring insect grubs inside branches, then extract the larvae using their elongated index fingers. They fill a similar ecological niche to woodpeckers, which are absent from Madagascar.

The front teeth are also used to pare away the hard wood of bamboos to get at the soft pith. A pair of aye-ayes was once described living in a coconut plantation. They came out in the evening and walked around the trees, each selecting a suitable nut. They used their incisor teeth to sink a hole, then used the middle finger to extract the milk and pulp, poking it in, then licking off the adhering mush. Aye-ayes use the same trick for drinking water, sweeping the long finger to the mouth at about 40 strokes a minute.

Breeding

Aye-ayes give birth to a single young, born in a large, spherical nest of leaves about 20 inches (50 cm) in diameter with an opening in one side. The nest is built by the female in a hollow tree or in the crotch of a branch. In some reports the young are born in October or November, other records say February and March. Nothing is known of mating, gestation or whether or not the male helps in the rearing of the young.

Conservation

In 1933 the aye-aye was thought to be extinct, but it was rediscovered in 1957 by Professor J.J. Petter. However, by 1966 it was considered that less than a dozen aye-ayes were left. Then the island of Nosy Mangabe, just off Madagascar, was declared a reserve and nine aye-ayes were introduced there. By 1976 there seemed to be only two or three left on the island. In 1989 there were only 10 individuals in captivity, but they do not breed successfully outside their natural habitat. To this day the aye-aye is considered to be critically endangered but it is still present in parts of Madagascar and is protected by law.

AYE-AYE

CLASS **Mammalia**

ORDER **Primates**

FAMILY **Daubentoniidae**

GENUS AND SPECIES *Daubentonia madagascariensis*

WEIGHT
Up to 6½ lb. (2.9 kg)

LENGTH
Head and body: about 16 in. (40 cm); tail: 22–24 in. (55–60 cm)

DISTINCTIVE FEATURES
Largest nocturnal primate; rounded face; large eyes and ears; coarse, shaggy gray or black coat; long, bushy tail; protruding, squirrel-like teeth; elongated index (middle) fingers

DIET
Insect larvae, fruits, bamboo shoots and tree sap

BREEDING
Age at first breeding: 2 years or more; breeding season: not known; gestation period: not known; number of young: 1; breeding interval: 2 or 3 years

LIFE SPAN
Up to 23 years in captivity

HABITAT
Forest, mangroves and bamboo thickets; occasionally cultivated areas such as coconut plantations

DISTRIBUTION
Restricted to isolated areas of northwest and northeast Madagascar; introduced to Nosy Mangabe Island, off Madagascar

STATUS
Critically endangered; thought to be one of rarest primates in world

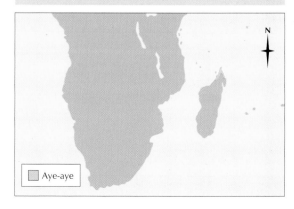

Aye-aye

BABBLER

A VARIETY OF BIRDS IS grouped together under the name babbler. These include about 260 to 279 species in all, some of which bear little resemblance to the others. There is little difference between males and females.

In general, babblers are poor fliers with short, rounded wings and fairly long tails. They live near the ground, feeding on insects and berries. Some species are drab, but others have brilliant markings. The smallest is the size of a wren, the largest the size of a crow. Their name comes from their loud and varied calls. Some species perform chorus duets, others antiphonal duets, in which a bird repeats the call of another.

The babblers are divided into seven major groupings: jungle babblers, Pellorneini; scimitar babblers and wren babblers, Pomatorhinini; tit babblers, Timaliini; song babblers, Turdoidini; parrotbills, Paradoxornithinae; wrentits and their relatives, Chamaeinae; and rockfowls, Picathartinae. Not all authorities agree that the last three should be classed as babblers, but there seems to be no other obvious place for them.

A well known member of the song babblers is the red-billed leiothrix, *Leithrix lutea,* or Pekin robin, as it is more popularly called. It is much like a European robin in build, but slightly larger. Its plumage is olive green, with red and yellow markings on the wings and throat. Its bright colors and attractive song have made it a popular cage bird. Despite its name, the Pekin robin comes from the Himalayan region of India and southern China. The closely related and slightly larger silver-eared mesia, *L. argentauris,* of the Himalayas and Southeast Asia, is also popular as a cage bird. It is more brilliantly colored than the Pekin robin, with a black head and silver gray patches on the ears.

A predominantly Asian family

Babblers are mainly inhabitants of the Old World. They are abundant in India, China and Southeast Asia, and occur as far away as New Guinea, Australia and Africa, including Madagascar. The only American babbler is the wrentit, *Chamaea fasciata.* A small, perky brown bird that frequently cocks its long tail, it lives in brushland on the Pacific coast from Oregon to Baja California.

Babblers occur in a wide range of habitats, from tropical rain forest to arid scrub. The largest babblers tend to feed on the ground, while smaller species forage in the canopy. The greatest diversity of species occurs in the Oriental fauna region, where 139 species are present.

Babblers are gregarious birds. Outside the breeding season, they congregate in flocks, often of several species, numbering about 3 to 30 birds. They regularly form clumps, huddling together.

The lowland mouse babbler, Crateroscelis murina, *is one of the smaller babblers. It displays the sandy buff coloring on its chest that is characteristic of many babbler species.*

Rufous babblers, Pomatostomus isidorei, gathered around a forest pool in New Guinea. This species has the long tail typical of most babblers.

They also preen each other. In some laughing thrushes and jungle babblers the preening is hierarchical, with senior members preening the newer additions. This helps to keep the group together, as does their continual babbling song, which informs each member of the flock where the others are as they move through dense forest.

Breeding strategies

At the start of the breeding season certain flocks split up into pairs that spread out and take up territories, defending them against other pairs. In some species, pairs remain together in groups throughout the breeding season and communal breeding may take place. Babblers such as the silver-eared mesias, yuhinas and jay thrushes will share a nest between more than one pair and build the nest together. The young may be fed by any member of the group. The nests are built near the ground, well hidden by vegetation. Some species build open-cup nests, while others build domed nests with entrances in the side.

A feature of many nests is the delicacy of the materials used: lichens, spiders' webs and the skeletons of leaves. Spiders' web is ideal for nest-building because it is sticky. The babblers use it to weave the other materials together.

The elusive rockfowl

There are two species of Picathartes, or rockfowl: gray-necked and white-necked. Both are fairly large and have brilliantly colored bald heads. On their discovery in the early 19th century they were attributed to the crow family and called bald crows. A century later they were moved to the starling family. Only in 1951 was it suggested that they be classed as babblers. Such uncertainty

COMMON BABBLER

CLASS	**Aves**
ORDER	**Passeriformes**
FAMILY	**Muscicapidae**
TRIBE	**Song babblers, Turdoidini**
GENUS AND SPECIES	***Turdoides caudatus***

WEIGHT
1⅛–1¼ oz. (35–40 g)

LENGTH
Head to tail: about 9 in. (23 cm); wingspan: 8–9½ in. (21–24 cm)

DISTINCTIVE FEATURES
Sandy buff plumage; strongly decurved bill; long, rounded tail

DIET
Mainly insects, fruits and seeds

BREEDING
Age at first breeding: 1–2 years; breeding season: all year, peaking March–July; number of eggs: 3 to 5; incubation period: about 13 days; fledging period: about 11–12 days; breeding interval: 1 year, but only 1 or 2 adult females in each group breed each year

LIFE SPAN
Probably up to 15 years

HABITAT
Semidesert areas, dry scrub and sandy plains; also cultivated trees, orchards and gardens

DISTRIBUTION
Southern Iran east to most of India

STATUS
Generally common

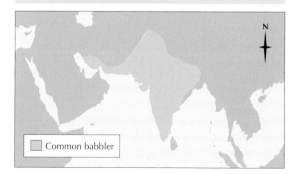

Common babbler

about these birds is not surprising, because they were known in Europe from only a few skins seen in museums. The grey-necked rockfowl, in particular, has rarely been seen alive, probably because it inhabits dense undergrowth. Unlike other babblers it builds its nest on rock faces.

BABOON

BABOONS HAVE RECEIVED considerable attention from zoologists over the years, one reason being that they often live in the open where they can be easily studied. Such research is important because baboons are monkeys that have forsaken a life in the trees, and so may give us clues about how our ancestors made a similar move. Both must have faced the same problems of getting food and guarding against danger.

Baboons belong to the family Cercopithecidae, the Old World monkeys. They inhabit most parts of Africa, where they live in family groups, called troops. Most are smaller than chimpanzees and gorillas, ranging from 2 to 4 feet (0.6–1.2 m) in head and body length. They have long, doglike muzzles and naked faces and the dominant males often have large teeth. Their tails, 1½–2⅓ feet (46–71 cm) in length, are usually held in a characteristic inverted "U" shape. One species, the drill, *Papio leucophaeus*, is actually somewhat larger, of a similar size to the chimpanzee. There is some debate as to whether the drill and the mandrill, *P. sphinx*, are in fact baboons, genus *Papio*, or belong to another genus, *Mandrillus*. These two species are discussed elsewhere.

Wide-ranging baboons

The savanna baboon, *P. cynocephalus*, lives in eastern, central and southern Africa. It is also known as the chacma, olive or yellow baboon, depending on its different forms or subspecies. The Guinea baboon, *P. papio*, meanwhile, is found in west-central Africa. The gelada baboon, *Theropithecus gelada*, is classified in a separate genus. It is confined to the mountains of Ethiopia, while the hamadryas or sacred baboon, *P. hamadryas*, is found farther north in Saudi Arabia, Sudan, Ethiopia and Somalia. Hamadryas baboons are now extinct in Egypt, but they, along with the other species, remain common throughout most of the rest of their range. Baboons usually inhabit savanna, rocky open country and bush, and occur in some mountainous regions. They are also sometimes found in woodland and on agricultural land.

Social organization

It is the social structure of baboons, the organization of the troops, that has attracted the interest of ethologists (scientists who study animal behavior). Each troop is a family unit, with all mating taking place between the members of that group. Typically the troop consists of older males, juveniles, females and babies. In a small troop there may be no more than one male with two or three females and their young, but large troops may number up to 50 animals. In addition, several troops may gang together to form herds.

The troop is a discreet unit and the members never wander far from each other. They have a definite range of countryside over which they wander, searching for food. At first sight a troop of baboons appears to have little order, but with close observation, a definite social structure can be seen. In the lead come some of the smaller males, followed by females and juveniles. In the middle of the troop there are the females carrying babies, the youngest juveniles and the older, dominant males. More females and young males bring up the rear. The advantage of this formation is that the females and babies are protected from all sides. Moreover, when danger threatens, the females and the very young start to

A subordinate hamadryas baboon grooming a large adult male. Mutual grooming not only keeps the fur clean, but also promotes good relations within a troop.

Chacma baboon, a subspecies of the savanna baboon, Hwange National Park, Zimbabwe.

BABOONS

CLASS	**Mammalia**
ORDER	**Primates**
FAMILY	**Cercopithecidae**
GENUS	***Papio** and **Theropithecus***
SPECIES	**6, including gelada baboon, *Theropithecus gelada*; Guinea baboon, *Papio papio*; savanna baboon, *P. cynocephalus*; and hamadryas baboon, *P. hamadryas***

WEIGHT
30–90 lb. (13.5–41 kg)

LENGTH
Head and body: 2–4 ft. (0.6–1.2 m); tail: 1½–2⅓ ft. (45–70 cm)

DISTINCTIVE FEATURES
Long tail; naked face with doglike muzzle; large teeth (dominant male only)

DIET
Fruits, seeds, nuts, tubers, shoots, buds, crops, insects, lizards, bird eggs and small mammals

BREEDING
Age at first breeding: 7–10 years (male); 5 years (female); breeding season: all year or seasonal; number of young: usually 1; gestation period: 180 days

LIFE SPAN
Up to 45 years in captivity

HABITAT
Savanna, scrub, bush and farmland; also mountains (especially gelada baboon)

DISTRIBUTION
Most of sub-Saharan Africa and some mountainous areas of Sahara; Saudi Arabia and Yemen (hamadryas baboon only)

STATUS
Generally common; hamadryas and Guinea baboons becoming rarer

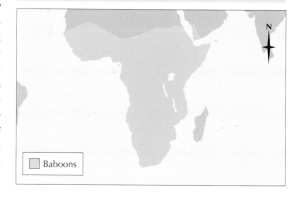

Baboons

flee first. The males move away more slowly, so they become congregated between the source of danger and the mothers and babies.

Baboon families appear to live amicably, the top male having no need to force the others to stay with him. Nor are there many signs of violence, except among the younger males. One behavioral trait that assists in holding the troop together is mutual grooming. When the baboons are resting, or feeding quietly, they gather in small groups to groom each other's fur. This keeps the fur clean, but more important, it promotes harmony between individuals. The grooming clusters usually form around a dominant male but other clusters also form, for example when the baboons gather around a female with a newborn infant.

To what extent there is a stronger bond than this between the members of a troop is difficult to say. Certainly it is essential for them to keep together as a protection against predators. However, according to some observers, if a baboon is sick or injured its fellows will show concern for it, ensuring that it is not left behind when the troop is on the move. Other observers report no such consideration. Presumably the presence or absence of such compassionate behavior depends on circumstances, such as the degree of danger. The age of the injured baboon may also be important. The cry of a young one when hurt, for example, will bring the adults running to its aid.

Early warning system

Spending the day in the open and sleeping at night in trees, as baboons do, is a sound defense. The tactics of a troop when faced by a predator are reinforced by a warning system. While moving about and feeding, the baboons keep up a chorus of quiet grunts. On being disturbed, any one of them will give a shrill bark to alert other members of the troop. If a female gives the alarm, one of the males will then move away from the troop to keep the intruder under observation and will give a double bark whenever it moves.

The main predators of baboons are cats, principally lions and leopards. Cheetahs and servals may also try to prey on baboons, but with less success. The usual reaction to attack is for the baboons to make for trees or rocks where, on reaching safety, they bark defiance and even throw stones. The older males are more courageous and sometimes turn on their predators, several older males being more than a match for a lion or leopard.

Breeding

Both males and females reach sexual maturity at around 5 years of age, but males will not mate successfully until they are between 7 and 10 years. In some regions mating is seasonal, while in others baboons breed throughout the year. When not pregnant or nursing, the females come on heat for a week in each month. Within the troop there is a hierarchy, or pecking order, among the males. There is usually one dominant male but sometimes several lower ranking males gang up to boss the others in the troop. Moreover, all males are free to mate with receptive females. Pairings are temporary, but when the female is at the peak of estrus she is guarded by the dominant male. Later, younger males mate with her.

Precocious young

A single young is born after a gestation of 180 days; occasionally twins are born. The newborn baboon soon clings to the hair of its mother's chest. Within hours of its birth it must have a good enough grip to hang on, even when the mother leaps into the trees. As it grows older, the young baboon learns to ride on the mother's back, jockey fashion, and soon after this it will begin eating solid foods and leaving the safety of its mother's body. Its excursions become more and more adventurous until it meets other young baboons, and starts to play with them. However, if danger threatens all young baboons run straight to their respective mothers.

The play group of young baboons becomes a very important part of their life. It is here they learn the skills needed for later life, in the form of games such as chasing and mock-fighting.

Omnivorous diet

Baboons eat a wide variety of foods, both plant and animal, depending on season, local availability and the age of the baboon. Often they will take anything that is available at the time. Seeds, shoots, tubers, buds and fruits are all eaten in season, and grasshoppers, butterflies and lizards are also caught. Scorpions are sometimes taken, the baboons nipping off the stings with their fingers. Occasionally they feed on small mammals, and even larger mammals, such as hares or smaller monkeys, may be chased, several baboons joining in to catch them. Newborn Thomson's gazelles, eggs and fledgling birds have also been recorded as prey.

By tilling the soil, humans have also provided baboons with a plentiful supply of food, mainly in the form of fruits, and in many parts of Africa baboons have become a serious pest of commercial crops.

A study by John Hurrell Crook showed that the troop structure of gelada baboons in Ethiopia depended to some extent on the amount of food available. Where there was an abundance of food the troops were large and contained many males, but in the arid areas a troop would have only one male. This meant that the females, which needed a plentiful supply of food for pregnancy and milk production, did not have to compete with males for the sparse supply. Yet the one male in the troop was sufficient to ensure procreation.

A female yellow baboon with her young, Tsavo East National Park, Kenya. A baby baboon is able to cling to its mother and travel with the troop within hours of its birth.

BACKSWIMMER

Backswimmers do not have gills and so have to come to the water's surface to breathe.

A GROUP OF LARGE-EYED aquatic bugs living in ponds, lakes and canals, the backswimmers include the boat-flies, water boatmen, or wherry-men, and the backswimmers themselves. They are distributed worldwide.

A widespread genus is *Notonecta,* with species such as *N. hoffmanni, N. insulata* and, one of the commonest, *N. glauca.* This last can be seen in almost any stretch of sluggish water and is rather more than ½ inch (1.3 cm) in length. It is of a pale brownish color with a darker thorax and undersides. Other species vary from ⅖–⅗ inch (1–1.6 cm) in length in the adult stage. Backswimmers are especially conspicuous for their long, paddlelike hind legs and for their characteristic habit of resting upside down just beneath the surface of the water.

Aqualung divers

The colloquial name backswimmer is particularly appropriate for these insects because they do actually swim upside down. Their very long hind legs, almost twice the length of the other two pairs, are fringed by a series of fine hairs. With a few strokes of these oarlike legs the backswimmers can propel themselves through the water at a remarkably fast rate. Their wingcases join to form a ridge along the middle of the back and when one of these insects is seen resting, just beneath the surface with its long hind legs held out sidways, the impression is of a keeled rowing boat with a pair of oars over the sides.

Backswimmers are extremely wary and at the slightest hint of danger will swim below water. This needs great physical effort because they are very buoyant due to a bubble of air which is always carried pressed to the abdomen by a series of bristles.

Although fully aquatic, backswimmers do not have gills. They must therefore get their air supply from outside, and this they do by rising periodically to the surface and sticking the tip of the abdomen out of the water. There is a channel formed by hairs on each side of the abdomen. These are opened at the surface, allowing air to flow in, and are then closed again, trapping the air. These air bubbles are therefore in direct contact with the backswimmer's spiracles, or breathing holes, which are arranged along the sides of the abdomen. The spiracles are protected by a further fringe of hairs, which allows air in and keeps water out.

BACKSWIMMERS

PHYLUM	**Arthropoda**
CLASS	**Insecta**
FAMILY	**Notonectidae**
GENUS	***Notonecta*; others**
SPECIES	**Many, including *Notonecta glauca*, *N. hoffmanni*, *N. undulata* and *N. insulata***

ALTERNATIVE NAMES
Boat-fly; water boatman; wherry-man

LENGTH
Adult: ⅖–⅔ in. (1–1.6 cm), depending on species

DISTINCTIVE FEATURES
Very long, hairy hind legs used as oars to aid swimming; swim upside down, with back shaped like underside of a boat

DIET
Mainly other insects; also crustaceans; occasionally tadpoles and small fish

BREEDING
Variable. Eggs laid into or on surface of aquatic plants; eggs hatch into nymphs; growth proceeds by a series of molts; 1 or 2 generations per year.

LIFE SPAN
Varies with species and environment

HABITAT
Variety of freshwater habitats, including ponds, pools, lakes, canals and ditches

DISTRIBUTION
***N. undulata*: North and South America, including almost every western state of U.S. *N. insulata*: western North America from British Columbia south to New Mexico. *N. glauca*: widespread globally.**

STATUS
Abundant

Although perfectly adapted to life in the water and scarcely able to walk on land, backswimmers are strong fliers and can leave their natural element at any time.

Voracious feeders

Backswimmers are extremely voracious feeders and it is unwise to include them with other small forms of life in an aquarium. Mosquito and other fly larvae form a large part of their diet, but size alone does not always deter them. Large beetle larvae, tadpoles and even small fish are often attacked. The backswimmer's method of hunting is to hang motionless at the surface of the water, immediately swimming toward anything that catches its attention. It has excellent eyesight, but primarily it discovers its prey by a form of vibration-location. Certain hairs on the hind legs can pick up the vibrations caused by small animals swimming nearby. Only when the backswimmer is within a few inches of its prey do the eyes play their part in securing its capture. Having captured its prey, the backswimmer then plunges its sharp rostrum (mouthparts) into the body of the victim, pumping in a digestive fluid containing enzymes that rapidly break down the body tissues. The carcass is then held firmly by the prehensile forelimbs while the internal tissues, now made fluid, are sucked out.

Life cycle

Like all bugs, backswimmers undergo incomplete metamorphosis. That is, a nymph hatches from the egg, resembling the adult insect in form. Growth then proceeds by a series of molts. In *Notonecta* the female uses her ovipositor to insert batches of the elongated, oval eggs into the stems of aquatic plants, such as Canadian pondweed. In other species the eggs are laid on aquatic plants. The eggs hatch after several weeks. The nymphs, which are at first wingless, escape by means of the hole originally made by the female's ovipositor. By late summer the nymphs have become young adults. There might be one or two generations per year, depending on species and evironment.

Backswimmers fall prey to the predatory animals present around its habitat. These include waterfowl, frogs, toads and sometimes fish, such as trout and bass.

Young backswimmers, called nymphs, closely resemble adults in form. They gradually develop into the adult without the larval and pupal stages of higher insects.

BADGER

THERE ARE EIGHT SPECIES OF badgers in six different genera, most quite different to one another in size, habitat and coloration. The European species, the Eurasian badger, *Meles meles*, is the one mainly discussed here. It ranges right across Europe and Asia, from Ireland and the British Isles to China, yet it is one of the most elusive of mammals. It is nocturnal and so wary that it is rarely seen by casual observers.

All species of badgers are bearlike animals with a stocky bodies, most about 3 feet (90 cm) in length, with short tails and short but powerful legs. The forefeet are armed with strong claws and they walk on the soles of their feet like a bear. Both bears and badgers are members of the order Carnivora but they are in different families. Badgers are placed in the family Mustelidae, along with the otters, ermines, martens and weasels. A characteristic of the mustelids is that their footprints show five toes. In this way, and because of its long claws, a badger's footprints can easily be distinguished from those of a dog, which show only four toes.

Another characteristic of mustelids is the musk glands at the base of the tail. The best known animals that have musk glands are the skunks. They will squirt a strong-smelling fluid whenever they feel threatened. Ermines and weasels use their musk to mark objects as a sign of ownership, and badgers also use their scent glands to mark territory in this way. Badgers also emit musk when frightened or excited.

An American badger emerges from its hole, Montana, the United States. The American species is similar in size, appearance and some of its habits to the Eurasian badger.

At a distance, the Eurasian badger's coat looks gray, but the individual hairs are black and white. Most animals are lighter in color on the underside of the body, but the badger has a black belly and legs. The most striking feature of this badger, however, is its head. It is white with two broad, black stripes running from behind the ears almost to the tip of the muzzle. The Eurasian badger's small eyes are located within these black stripes, so they are fairly inconspicuous.

The American badger

The American badger, *Taxidea taxus*, unlike many other species, bears a strong resemblance to the Eurasian badger. It is similar in size and grayish in color, with dark legs. It also has the black-and-white face, although there is just one central, white stripe extending from its nose to its back. It is widespread in North America from southwestern Canada south to central Mexico.

Badgers of Southeast Asia

In Southeast Asia there are several other species of badgers. These include the hog badger, *Arctonyx collaris*, also called the sand or hog-nosed badger. It has very similar habits to its Eurasian relative, but can easily be distinguished by its naked, piglike snout, from which it gets its name, and its much longer tail.

There are also three species of ferret badgers, genus *Melogale*, also known as tree badgers or pahmi. They are found in the grasslands and forests of Southeast Asia and are brownish or blackish in color, with white markings on the face and throat and sometimes on the back.

There are two more species, known as stink badgers: the Malaysian stink badger, *Mydaus javanensis*, and the Palawan or Calamanian stink badger, *Suillotazus marchei*. These are rare inhabitants of Southeast Asia, the last found only on Palawan in the Philippines. The honey badger, *Mellivora capensis*, is discussed elsewhere.

Rarely seen

The Eurasian badger often lives surprisingly close to the center of large cities and is found in practically all European and Asian countries, from just south of the Arctic circle to the Mediterranean and the Himalayas. In the northern parts of its range it hibernates, but in other regions it may be active all winter.

Despite being relatively common, badgers are rarely seen because of their nocturnal habits. It is extremely rare for a badger to be seen away from its set (badger hole or burrow) during

EURASIAN BADGER

CLASS **Mammalia**

ORDER **Carnivora**

FAMILY **Mustelidae**

GENUS AND SPECIES *Meles meles*

ALTERNATIVE NAME
Brock (archiac)

WEIGHT
22–36 lb. (10–16 kg)

LENGTH
Head and body: up to 3 ft. (90 cm); tail: about 6 in. (15 cm)

DISTINCTIVE FEATURES
Black-and-white, stripy face; small, bearlike body with short tail; small eyes and ears; short, powerful legs; forefeet armed with large, powerful claws

DIET
Mainly earthworms; also snails, insects, fruits, acorns, grasses, bulbs, small mammals, frogs and carrion

BREEDING
Age at first breeding: 1–2 years; breeding season: mates in spring, delayed implantation to December; number of young: usually 2 or 3; gestation period: about 49 days; breeding interval: 1 year

LIFE SPAN
Up to 11 years

HABITAT
Farmland, woodland and mountains

DISTRIBUTION
Throughout central Eurasian region, from Ireland and Britain east to China

STATUS
Generally common, but declining locally in parts of range

Eurasian badger

daylight hours. During the long, dark nights of winter, badgers may not emerge at all. In addition, if there is any disturbance, suspicious sound or scent, the badger may remain underground for the entire night, and it will often fail to come out on bright, moonlit nights.

Set dwellers

Badger sets are easily distinguished from the dwellings of foxes or rabbits because of their large size and the mass of earth and stones that lies at the entrance. However, foxes and rabbits may take up residence in sets, sometimes even when the badgers are still in occupation.

Signs of the badgers' presence are unmistakable, however. They have regular, well-trodden paths leading from the set. These may be followed for some distance, often running to a stream or pond where the badgers habitually drink. Near the entrance to the set there are more definite signs of activity. Scratching posts indicate where a badger has stood on its hind legs and scratched the trunk of a tree with its forepaws. Around the mouth of the hole, and along the paths leading to it, fresh vegetation is often strewn. This usually comprises bracken, flowers and other plants the animals have collected for bedding. The badgers gather vegetation in their forelegs and shuffle backward, leaving a trail of plants.

Badgers have a reputation for being especially clean animals, largely because they frequently change their bedding and also dig latrines, shallow pits that can be found within 60 feet (18 m) or so of the set. The favorite sites for sets, which may be vast underground systems with many entrances, are in woods, preferably those bordered by pastures. The badger seems to prefer sandy soil.

Although common throughout most of its range, the Eurasian badger is rarely seen because of its nocturnal lifestyle. It is also wary of humans, probably as a result of a long history of persecution.

Cubs born underground

The boar (male) and sow (female) badgers are thought to pair for life. In the case of the Eurasian badger, mating takes place in spring but, because of delayed implantation (in which implantation of the embryos in the womb is delayed for some time), the embryos do not begin to develop until about December. There may be one to five cubs, but normally two or three are born during February. At birth they measure no more than 5 inches (13 cm) in total length.

For the first 6–8 weeks of their life, the young stay underground, then make their first, tentative visits to the outside world. They stay out for only a very short time at first, scampering back to safety at the slightest alarm. After a week or so they become bolder and start exploring the neighborhood. Later the cubs are taken out by their mother to learn to feed themselves and they eventually leave their parents around October.

Main food is earthworms

A Eurasian badger at the entrance to its set at sunset. This species forages at night, mainly for earthworms, but it also takes other small prey, fruits and carrion.

As would be expected of a carnivore, an examination of a badger's skull suggests an animal equipped for attacking and consuming large prey. The teeth are strong and there are long ridges around the hinges of the jaws that prevent them from being dislocated. Despite this, the Eurasian badger lives on a wide variety of soft food, and earthworms make up the majority of its diet. Other animal matter includes mice, voles, moles, frogs, snails, beetles and even hedgehogs and wasps, while windfall apples, bulbs, acorns, blackberries and grass are also eaten. Cereal crops might suffer when badgers flatten areas to get to the ears. Badgers occasionally kill poultry, but this is not typical and will occur only if there is a scarcity of their normal food.

The American badger feeds mainly on small mammals such as rabbits, prairie dogs, ground squirrels, mice and voles. It is well known for being able to burrow rapidly in order to reach these animals. It will also eat ground-nesting birds and their eggs, and sometimes takes snakes.

Predators

Badgers have few natural predators, but humans have in past times trapped them for their fur, or to provide sport by badger-baiting with dogs. Farmers sometimes kill badgers because they think they are damaging their crops or taking their poultry. Badgers are also persecuted because their burrowing may be hazardous to horses or cattle. Although the species is protected, official culls of the Eurasian badger have been organized in Britain because it is thought that these animals might pass tuberculosis to cattle. There is little scientific evidence to support this theory.

BALD EAGLE

ONE OF THE GROUP KNOWN as sea eagles, the bald eagle of North America is specialized in hunting fish. Bald eagle is an inaccurate name, but an impression of baldness is given by the bird's snow-white head and neck, contrasting with the brownish black plumage of the rest of its body. Its tail is also white and it has a large, yellow bill and yellow eyes and legs. Adults reach up to 3⅗ feet (1.1 m) in length, with a wingspan of perhaps 11½ feet (3.5 m). Bald eagles do not get their white feathers until they are 4 or 5 years old and might be mistaken for golden eagles up until this point. The remainder of the juvenile's plumage is lighter and more mottled than that of the adult.

The bald eagle has become one of the most familiar of eagles from its use on the seal of the United States of America.

No longer endangered

At one time bald eagles bred extensively throughout North America. Formerly, the northern boundary of their range extended east from Bering Island and Alaska, following a line down to the south of Hudson Bay in northern Canada, then back to Labrador in Newfoundland. The southern boundary extended from lower California in the west to Florida in the east. However, extensive use of the pesticide DDT (which results in sterility) in the 1960s, coupled with shooting and habitat destruction, caused serious declines in the 48 state populations outside Alaska. By the end of the 1960s, as few as 417 nesting pairs remained outside the Alaskan stronghold, but this number has since climbed back to almost 5,800 pairs. In 1999, the bald eagle was removed from the endangered species list of the U.S. Fish and Wildlife Service. There are two subspecies: the northern and southern bald eagles. The differences between these are not great and they are difficult to tell apart. The southern subspecies is now confined to South Carolina, Florida, the states around the Gulf of Mexico and Texas.

Like many other raptors (birds of prey), bald eagles spend much of their time perching motionless, taking in every movement within their wide range of vision. They are generally found close to inland or coastal waters, but may be seen in more mountainous regions during migration. They are solitary birds, occasionally seen roosting together in a tree but otherwise being found in large numbers only where there is a plentiful source of food. Up to 4,000 have been recorded congregating along a 10-mile (16-km) stretch of the Chilkat River, Alaska, during winter, attracted there by abundant salmon.

Mainly fish eaters

Bald eagles take many kinds of prey, and will also feed on carrion. For most of the year, however, fish provide the bulk of their diet, including dead fish that have been washed ashore.

A bald eagle in flight. Adult bald eagles have a wingspan of up to 11½ feet (3.5 m).

A bald eagle with its catch, Kachemak Bay, Alaska. Bald eagles are skilled fishers, but often prefer to rob ospreys of their prey.

The bald eagle's method of catching live fish varies. It is highly adaptable and able to take advantage of almost any opportunity for feeding. Sometimes it fishes in the same manner as the osprey, circling over the water, hovering on rapid wingbeats as it spies its quarry, then dropping onto its prey. At other times it will search a lake by flying leisurely just above the surface, or will wait patiently in a nearby tree until it sees a fish.

A more spectacular and regular habit of the bald eagle is that of systematically robbing ospreys. The eagle, sitting on a suitable perch near the osprey's hunting ground, waits for the other bird to appear laden with its prey. On seeing the eagle, the osprey attempts to escape but the eagle, not being weighed down by a fish, soon overtakes it and harries the bird until it is forced to drop its catch and flee. The eagle then hurtles down to retrieve the fish. Occasionally the bald eagle will even snatch prey away from the osprey using its talons.

When fish are not so plentiful, bald eagles turn to other prey such as rabbits, squirrels, waterfowl, shorebirds, puffins and rats. Larger mammals are sometimes caught and the remains of young caribou, mule deer and lambs have been found around eagle nests. However, it is likely that most of these died from other causes

BALD EAGLE

CLASS	**Aves**
ORDER	**Falconiformes**
FAMILY	**Accipitridae**
GENUS AND SPECIES	***Haliaeetus leucocephalus***

ALTERNATIVE NAMES
American eagle; white-headed sea eagle

WEIGHT
7¾–11 lb. (3.5–5 kg)

LENGTH
Head to tail: 2⅘–3⅗ ft. (0.85–1.1 m); wingspan: 6½–11½ ft. (2–3.5 m)

DISTINCTIVE FEATURES
Adult: snow-white head and tail; brownish black body; yellow eyes, bill and legs. Juvenile: lacks white areas; lighter and more mottled elsewhere.

DIET
Mainly fish; also waterfowl, shorebirds, small mammals and carrion

BREEDING
Age at first breeding: 4–5 years; breeding season: October–February in southern states, later farther north; number of eggs: 2 or 3; incubation period: 35 days; fledging period: 72–75 days; breeding interval: 1 year

LIFE SPAN
Up to 20 years

HABITAT
Usually close to inland or coastal waters; also in mountains during migration

DISTRIBUTION
Throughout most of U.S. and Canada, from Alaska south to Florida and California

STATUS
Locally common in parts of Alaska and northern Canada; uncommon elsewhere

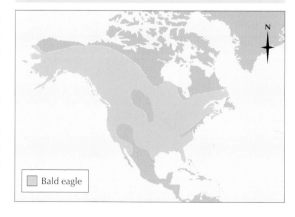

Bald eagle

and were picked up by the birds as carcasses. Nevertheless, some farmers assert that bald eagles take sheep, and will shoot them.

The methods employed to catch waterfowl also show the bald eagle's adaptability. Most frequently waterfowl are pounced upon, on land or in the water, and carried off, but the eagle is also capable of forcing them down while in flight. Ducks and other diving birds may be forced by the eagle to dive again and again until exhausted, and are then easily captured. Even more remarkable is a method of catching geese on the wing. The eagle dives under the goose, rolls over and sinks its talons into its breast.

Lifelong partnership

Bald eagles mate for life and also cling tenaciously to their nesting site. The nest is used year after year, and the eagles will stay put even if humans begin to develop the area.

In Florida the breeding season begins very early, in October or November. The reason for this early nesting is thought to be the abundance of waterfowl in winter, which can provide ample food for the growing eaglets. It may also be an advantage for the chicks to have grown an insulating layer of feathers to keep them cool before the weather becomes too hot. In other southern states breeding takes place between November and February, but it starts later farther north.

In the south the nests are in lone trees, usually some 45–70 feet (14–21 m) above the ground, pines being preferred. In Alaska, beyond the tree line, the eagles build their nests on rocky cliffs or pinnacles. Being used many years running, with additions every year, bald eagle nests become impressive edifices. They are made up of a great pile of sticks, some of them 6 feet (1.8 m) long, with a mass of weeds, stubble and adhering earth in the middle. During the season the whole of the nest becomes trampled flat, so each year a new, 1-foot (30-cm) high rampart is built.

Only one chick survives

The normal clutch is of two or three eggs. They are white or pale blue, and 2½ inches (6.5 cm) long. Incubation takes 35 days. Both parents take part in brooding and, later, in feeding the chicks. It is very rare for three to survive, and usually only one chick fledges, having bullied its nest mates out of their share of the food.

As the eaglets grow they are guarded only in bad weather, when one parent will stand over them with wings spread. Gradually they are taught to tear up their own food, and as their feathers develop they begin flying exercises, flapping their wings and even flying up a few feet. Eventually the parents lure them into proper flight with food as a reward, but they still return to the nest until the end of the summer, when their parents drive them away. Generally the fledging period is between 72 and 75 days and the young birds reach sexual maturity at 4 or 5 years.

Mobbed by other birds

Apart from humans bald eagles have no real predators, but several birds are known to mob them. Ospreys sometimes hit back as they are being robbed of their fish, and crows will repeatedly fly at raptors in order to protect their nests, eggs and chicks and themselves. They will harass the eagles, even landing on their backs and pecking at their heads. Small birds, such as kinglets and gnatcatchers, fearlessly pester any eagle that comes too near their nests.

The national emblem

On June 20th, 1782, the citizens of the newly independent United States of America adopted the bald or American eagle as a national emblem. At first the heraldic artists depicted a bird that could have been a member of any of the larger species, but by 1902 the bird portrayed on the seal of the United States of America had assumed its proper white plumage on the head and tail.

Although eagles are traditionally used as emblems of power, the choice of the bald eagle came in for some criticism. Benjamin Franklin preferred the wild turkey and said the bald eagle "like those among men who live by sharping and robbing... is generally poor and often very lousy." This was an allusion to its robbing ospreys, and its habit of eating carrion was also held against it. Nevertheless, the bald eagle's noble appearance has preserved its status as America's national bird.

The bald eagle, the only eagle confined to North America, was adopted by the United States as a national emblem in 1782. Large numbers of bald eagles gather at favored feeding grounds during the winter.

BALTIMORE ORIOLE

The Baltimore oriole's nest is like a basket hanging from a branch, with the entrance near the top. Here a male feeds its young.

THE MELODIC, FLUTELIKE song of the Baltimore oriole is one of the welcome sounds of spring in many parts of North America. With its black hood and bright orange underparts, rump and tail, along with its black-and-white wings, the Baltimore oriole is also one of the most attractive new arrivals each April and May. The species is found right across the United States east of the Rockies, and from northwestern British Columbia to Nova Scotia in Canada.

Like the other New World orioles, the Baltimore oriole is a member of the Icteridae family, the 104 members of which include species as diverse as the eastern meadowlark, *Sturnella magna*, and the brown-headed cowbird, *Molothrus ater*. Seven species of orioles are found in the United States and Canada, the Baltimore oriole being the most common of these.

The origin of the Baltimore oriole's name goes back to early colonial times. When George Calvert, an early English colonist and the first Baron of Baltimore, saw the species he was so impressed with its bright colors that he adopted it for his coat of arms. The bird was later named after him and is now the state bird of Maryland.

Long-distance migrant

The Baltimore oriole migrates long distances between its winter quarters and the areas in which it breeds. The species winters in Central America and South America, leaving there between mid-March and mid-April. It reaches the southern United States early in April and the Canadian border late in the month. By late May it has reached its most northerly breeding grounds in British Columbia. The males arrive about a week earlier than the females.

Research done on birds killed on migration in the autumn suggests that adult males migrate on a narrower path than either adult females or young birds. Juveniles begin to leave the breeding grounds as early as the first half of July, with migration gathering pace during August. Most have left by September.

Small numbers of Baltimore orioles winter in southern California and along the east coast of the United States from Virginia to Florida, while a few birds have braved the winter conditions as far north as Ontario, Canada. However, the vast majority spends the winter months from southeastern Mexico south to northern Colombia and northern Venezuela. Most of these migrant birds move southeast through the Caribbean slope of Mexico in autumn, but in spring more of the returning birds make the sea crossing over the western Caribbean. However, the species is rarely found east of Jamaica.

Return year after year

The interesting thing about the Baltimore orioles that winter in the southeastern states is that birds return there year after year. This fact has been discovered by programs of trapping and banding. So site-faithful are some that one individual was trapped at the same place in six years out of seven. The trend for spending the winter in the United States seems to have been encouraged by an increase in the number of people putting out bird feeders. Some ringing recoveries are most peculiar. A bird ringed in Rhode Island in October 1963, for example, was found a month later many miles to the northeast, in Newfoundland. It should have migrated in the opposite direction! Several birds have also crossed the Atlantic Ocean to western Europe in the autumn.

Singing frenzy

When they reach the breeding grounds in spring, the male Baltimore orioles start up an almost continuous frenzy of singing. The interval between bursts of this flutelike song may be just 4 seconds. Once they find a mate, the singing tails off. By late May most of the birds that are still singing continuously are unpaired males fledged the previous summer. Rarely, females also sing, though their song is not as melodic.

While it is thought that virtually all females mate, including one-year-old birds, few males of this age are successful. So while the mating

BALTIMORE ORIOLE

CLASS	**Aves**
ORDER	**Passeriformes**
FAMILY	**Icteridae**
GENUS AND SPECIES	***Icterus galbula***

ALTERNATIVE NAME
Formerly known as northern oriole

WEIGHT
1–1⅔ oz. (25–47 g)

LENGTH
Head to tail: 8⅔ in. (22 cm)

DISTINCTIVE FEATURES:
Male: black head and back; bright orange rump; black wings with orange epaulet (shoulder patch) and wide, white wing bar; mainly orange tail. Female: duller orange below; browner wings.

DIET
Insects, especially caterpillars; also berries and sometimes flower nectar

BREEDING
Age at first breeding: 1–2 years; breeding season: eggs laid April–late June; number of eggs: usually 4 or 5; incubation period: 12–14 days; fledging period: 12–14 days; breeding interval: 1 year

LIFE SPAN
Up to 12 years

HABITAT
Open woodland, riverside trees and parks, including in suburban areas

DISTRIBUTION
Breeding: U.S. east of the Rockies and Canada from northwestern British Columbia to Nova Scotia. Winter: southeastern Mexico to Colombia and Venezuela.

STATUS
Common

Baltimore oriole ▢ breeding ▢ winter

system is broadly monogamous, males will also mate outside the established pairing. The technical name for this is opportunistic polygyny.

Hanging nests

The Baltimore oriole's nest is an interesting construction, a pendant basket hanging from a branch with the entrance at the top. A big problem for many North American songbirds is that of parasitism by cowbirds. These birds lay their eggs in the nests of other bird species, particularly smaller ones. Once hatched, the young cowbirds tend to have a competitive advantage over their hosts' nestlings, so the cowbirds prosper and the hosts suffer. The break-up of large tracts of forest by roads and other development has enabled cowbirds to reach far more birds' nests than previously, as they are reluctant to venture more than ⅗ mile (1 km) away from the forest margin.

The fact that Baltimore orioles have hanging nests does not prevent cowbirds laying their own eggs in them, but unlike many species the orioles can recognize the alien eggs, and they get short shrift. The oriole stabs the cowbird eggs with its bill and throws them from the nest. This ability to deal with the parasite is doubtless one reason why the Baltimore oriole population has remained stable in most areas, unlike many other songbirds.

Caterpillar lovers

Baltimore orioles have also benefited from having a varied diet. Although they have a particular fondness for caterpillars taken from leaves, and will take even very hairy ones that other birds avoid, they also consume berries, nectar and fruits. One was seen to kill a ruby-throated hummingbird, presumably to eat, but this is exceptional behavior.

An immature male Baltimore oriole. While all females mate, even at 1 year of age, males are rarely successful at this age. As a result, the courtship song of unpaired young males may still be heard into late May.

BANDED ANTEATER

Formerly widespread throughout southern Australia, the banded anteater is now mainly restricted to the southwestern part of the country.

SOMETIMES CALLED THE marsupial anteater, this termite-eating animal might be better known as the numbat, the name given to it by the Aborigines. Still, it is now generally referred to as the banded anteater and we refer to it by that name here. Although it has in the past been said to eat both termites and ants, it now seems certain that it eats mainly termites. It is an official emblem of the state of Western Australia.

The banded anteater is nearly 1½ feet (46 cm) long, of which just under 7 inches (18 cm) is bushy tail. It is reddish yellow to chestnut red in color, with six or seven white bands around its body. Its stout body is flattened across the hindquarters and its muzzle is long and tapering. There is a black line running from its ear through the eyeline to the tip of its nose on either side of its face. Its skull is long and flattened, and it has 52 teeth, more than in any other mammal except some of the whales, dolphins and porpoises.

Sole member of its family

Previously placed in the family Dasyuridae, the banded anteater or numbat is now generally regarded as being the sole member of the family Myrmecobiidae. It used to be thought there were two species, the southeastern or rusty banded anteater and the banded anteater. These are now treated as one species. It originally ranged from western New South Wales across south and central Australia to Western Australia. However, in recent years has been found only in southeastern Western Australia through northern South Australia to southwestern New South Wales. Its range seems to be diminishing as human settlements spread.

Termites help provide shelter

Although sometimes lethargic, the banded anteater will suddenly run in a series of bounds and readily climbs trees. Occasionally it will stand up on its hind legs to look around. Normally its tail is carried straight out behind or with an upward curve. In moments of excitement the animal curves its tail upward, or even over its back, and the tail hairs are fluffed out.

The banded anteater is found in open scrub, sometimes desert, and in eucalyptus woodland. When the eucalyptus trees shed their branches they are soon hollowed out by termites, and this marsupial uses them for shelter. It takes in leaves and grasses to form a nest. Away from the eucalyptus trees it will dig out short burrows for overnight shelter in cold weather.

Eats 7 million termites a year

The feeding habits of the banded anteater were first described by the Australian naturalist David Fleay, in 1952. He had one of these animals in captivity and offered it termites, several kinds of ants and their eggs, mealworms, beetles, insect grubs, earthworms, raw egg, bread and milk, honey and jam. However, it took only termites, swallowing the small ones whole, and chewing the larger species. Fleay found that it ate around 10,000 to 20,000 termites each day.

In the wild, the banded anteater licks up termites from the termite galleries in rotten wood. It uses its cylindrical, extensible tongue, flicking it out rapidly in all directions to a length of 4 inches (10 cm). Later studies have found that ants make up only 15 percent of the banded anteater's entire diet. This percentage tends to be ants that have invaded termite colonies. While searching for termites, it uses its long snout as a probe or else rips open termite-infested wood or colonies in the earth, using its sharp claws.

BANDED ANTEATER

CLASS	**Mammalia**
ORDER	**Marsupialia**
FAMILY	**Myrmecobiidae**
GENUS AND SPECIES	***Myrmecobius fasciatus***

ALTERNATIVE NAMES
Numbat; marsupial anteater

WEIGHT
10–16 oz. (275–450 g)

LENGTH
**Head and body: 6⅔–11 in. (17–28 cm);
tail: 5–6⅔ in. (13–17 cm)**

DISTINCTIVE FEATURES
**Stout body, flattened across hindquarters;
long, tapering snout; prominent ears;
reddish yellow or chestnut red coat
with 6 or 7 white stripes across back;
prominent dark cheek stripe; bushy tail**

DIET
Almost entirely termites; some ants

BREEDING
**Age at first breeding: not known; breeding
season: August–December; number of
young: 2 to 4; gestation period: not known;
breeding interval: 1 year**

LIFE SPAN
Up to 6 years in captivity

HABITAT
**Open scrub and eucalyptus woodland;
also desert**

DISTRIBUTION
Southern and southeastern Australia

STATUS
**Vulnerable; suspected population decline
of 20 percent since 1990**

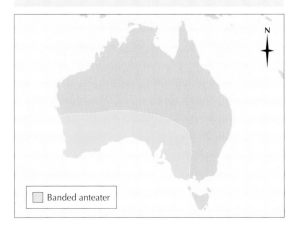

Banded anteater

Marsupial without a pouch

Although it is a marsupial, the banded anteater has no pouch. The female, which is markedly smaller than the male, has four teats surrounded by long, crimped hairs. Mating takes place between August and December, and two to four young are born, although the usual litter size is four. At first the babies merely cling to the teats with their mouths, but later they cling to the crimped hairs with their forefeet. They are carried by the mother in this manner for about 4 months or until the young have reached a certain size. At this point the mother digs a hole in the ground where she can leave the young while she goes out foraging.

Hiding out in hollow logs

Once inside a log, the banded anteater points its head toward the closed end of the log, tucks its tail under its body and closes the entrance with its broad rump. It can also swell its body so it fits into the hollow like a cork in a bottle. This is a trick used by a number of animals. The anteater's two possible natural predators, the carpet snake and a giant monitor lizard known as the goanna, cannot get to the animal when it is entrenched in this way. They must catch it in the open, perhaps surprising it while it is feeding.

Forty years ago there were fears that the banded anteater might be on the road to extinction, due, so it was thought, to its being killed by introduced foxes and domestic cats and dogs Later research suggested that it was the clearing of land, and especially bush fires, that constituted the greatest threat to this species. Not only is its range considerably less than it was, but population numbers are now thought to be down by about 20 percent on what they were in 1990.

*Banded anteaters
find shelter either in
short burrows or in
eucalyptus logs that
have been hollowed
out by termites, their
main food source.*

BANDICOOT

BANDICOOTS BELONG TO THE marsupials, the order of pouched mammals that includes the kangaroos, wallabies and koala. The distinctive feature of the marsupials is that their young are born when only partly developed. They crawl up the mother's fur and into the pouch, where they fasten onto the teats.

All bandicoots are ratlike in appearance, with long, pointed snouts, naked ears and hairy tails. On average they range between 8 inches and 1⅘ feet (20–56 cm) in length, with a 5–11-inch (13–28-cm) tail. The largest species is the giant bandicoot, *Peroryctes broadbeuti*, which can be up to 2 feet (60 cm) in length with a 13-inch (33-cm) tail. It weighs up to 10 pounds (4.5 kg). The bandicoots' fur is made up of two distinct coats, an outer coat of long, coarse hairs covering a soft underfur. The toes of all four feet bear sharp claws, and in some species the second and third toes of the hind feet are joined in a common envelope of skin, like a mitten, with only the last joint of each toe and the claws showing. These combined toes are used for grooming.

The hind feet of the now extinct pig-footed bandicoot, *Chaeropus ecaudatus*, underwent an even greater change. Not only were the second and third toes united, the first had disappeared, the fifth was hardly visible, while the fourth toe bore a large claw. The pig-footed bandicoot was

so named because its front feet evolved like those of the cloven-hoofed animals such as pigs. These changes to its structure were adaptations for running.

Many species now extinct

A total of 21 species of bandicoots have been described by zoologists to date, but this includes a number of species that are now extinct. Surviving species are spread over Australia, Tasmania, New Guinea and the surrounding region. This area is the stronghold of the marsupials. Elsewhere they have been superseded by the placental, or true, mammals. Apart from the opossums of North and South America, the marsupials now survive only in the Australasian region, which was isolated by the sea before the true mammals had spread that far. As a result, the marsupials were able to live without competition until Europeans appeared with their domestic animals. Nowadays the forests have been cut down, so destroying the marsupials' homes, while introduced cats, dogs, foxes and rats prey on them. Some marsupials are now extinct while others have become very rare.

Bandicoots, in particular, have suffered from indiscriminate shooting and rabbit extermination programs, for snares and poisons are not selective. Several species that were once common throughout Australia are now extinct. These include the pig-footed bandicoot, the desert bandicoot, *P. eremiana*, and the lesser bilby, *Macrotis leucura*. Others, such as the western barred bandicoot, *Perameles bougainville*, are endangered and found only in isolated parts of their former range.

Current threats

The main threat to the bandicoots remains habitat loss as a result of human activities such as cattle ranching, as mentioned above. Natural predators include the monitor lizards. The aborigines, the indigenous people of Australia, also hunt them for food. In addition, these people use bundles of the black-and-white tails of the greater bilby or greater rabbit-eared bandicoot, *Macrotis lagotis,* as personal ornaments.

Nests and burrows

Bandicoots are nocturnal, most retiring during the day to nests of grass and sticks in the undergrowth or under leaves. The nests have no regular entrance and the bandicoots force their way in and out anywhere, closing the gap behind them. The greater bilby, on the other hand, digs burrows. It has long, pointed ears, a rather bushy

An eastern barred bandicoot, Perameles gunnii, *scavenging among refuse. Once common across Australia, many species of bandicoots are now extinct, endangered or vulnerable.*

BANK VOLE

CLASS	**Mammalia**
ORDER	**Rodentia**
FAMILY	**Muridae**
GENUS AND SPECIES	***Clethrionomys glareolus***

ALTERNATIVE NAMES
Wood vole; bank mouse; red mouse

WEIGHT
1–1½ oz. (28–43 g)

LENGTH
Head and body: 3½–4 in. (9–10 cm);
tail: 1½–2 in. (3.8–5 cm)

DISTINCTIVE FEATURES
Small vole with blunt nose; reddish back;
underparts are whitish, varying from gray to
yellowish or buff; black upperside to tail;
ears almost hidden in fur

DIET
Mainly fruits, seeds, roots and bulbs;
also nuts, grain, fungi, moss and lichen;
occasionally insects, snails and other
invertebrates

BREEDING
Age at first breeding: female matures at 1
month, but breeding usually suppressed in
first year; breeding season: April–October,
sometimes all year; number of young: usually
3 or 4; gestation period: 18 days; breeding
interval: up to 4 or 5 litters per year

LIFE SPAN
Up to 18 months, average 2–3 months

HABITAT
Deciduous woodland, grassland and
hedgerows; also conifer forests and
plantations in northern regions

DISTRIBUTION
Much of Europe and parts of Central Asia

STATUS
Abundant

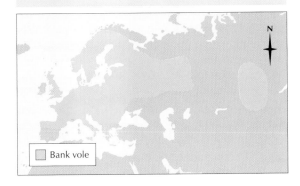

Bank vole

abundance, in which the population of these animals increases at a great rate. Toward the end of each cycle, the population suddenly decreases. After the decrease, a rapid reproductive rate allows the bank vole quickly to regain its numbers, but this variation in numbers is by no means as spectacular as in lemmings.

Bank voles are an important part of the diet of woodland predators such as owls, kestrels and weasels.

Hunted by woodland predators

Bank voles are taken in large numbers by many predators, especially those that hunt in woodland. Barn owls, tawny owls, weasels and kestrels are the bank vole's main predators, and changes in numbers of kestrels, in particular, appear to parallel the changes in numbers of bank voles. As the bank voles increase, so providing a superabundance of food, the kestrel numbers also increase markedly.

Vegetarian feeders

Bank voles are mainly herbivorous, feeding on roots, bulbs, fruits and seeds, along with nuts, berries, the grain of wheat and barley and the seeds of smaller grasses. Insects and their larvae make up a small part of the vole's diet, while snails and other small animals may also be eaten. In spring, bank voles climb rose and hawthorn bushes to feed on the new leaves, and in autumn they eat the hips and haws. Bank voles have long been thought of as pests by foresters because they occasionally strip bark from small trees, usually larch or elder. They also eat and bury seeds, and take buds and new shoots. This might reduce germination of the next generation of tree saplings, but it is difficult to assess to what extent woodlands are affected in this way by voles.

BARBARY APE

monkeys have tails while apes do not. The barbary ape weighs approximately 25–33 pounds (11–15 kg) and is 2–2⅓ feet (60–70 cm) long. Its coat is thick, coarse and yellowish brown in color, and it has a naked, pale pink face.

Survivors of a larger population?

Apart from Gibraltar, the barbary ape is found in a few parts of Morocco and Algeria in North Africa. On Gibraltar it lives on the rocky cliffs and scrub, but in North Africa it is found in high forests in the Atlas Mountains. Not only is it unique as a monkey in Europe, the barbary ape is also the only macaque found in Africa. All other macaques are found in Asia. It is presumed that either the barbary ape or its immediate ancestors once spread across North Africa as well as Europe, but due to a change in climate, the advance of humans or some other reason, their numbers dwindled, leaving just a few, scattered populations.

Whether the barbary apes of Gibraltar really belong there and are the last survivors of a population the fossil remains of which are to be found across Europe, or whether they were brought across the Strait of Gibraltar (the passage between Spain and Africa) by humans, are questions that are debated by zoologists. One thing is certain, they did not make their way across the Strait by a tunnel, as legend once had it. Whatever their origin, the apes now live a semi-domesticated life. The present population on the island is very small. Having dropped to seven in 1942 it rose to 30 by 1955, mainly through the custodians importing more monkeys from North Africa. Today the Gibraltar monkeys make up a small fraction of the worldwide population of barbary apes, around 75 percent of which are found in Morocco.

Barbary apes are diurnal (day-active), spending the night in holes or crevices in rocks. During the day they roam in large bands about the rocks and in the trees, for although terrestrial, they are good climbers.

Omnivorous feeders

The barbary ape feeds on a wide variety of foods, including leaves, pine cones and fruits, together with insects, scorpions and any other small creatures. In some parts of North Africa it is a serious pest, raiding and ruining crops.

On the Rock of Gibraltar barbary apes are less of a nuisance, but even in their small numbers the monkeys sometimes cause trouble when they descend on the town in search of food. Not only will they steal food, but their apparently insatiable

Legend has it that if the small colony of barbary apes leaves the Rock of Gibraltar, then the British will lose control of the territory.

Two attributes make the barbary ape unique. It is the only primate, the group of animals to which monkeys, apes and humans belong, that lives in Europe today (apart from humans, of course). A small population exists on the British-held Rock of Gibraltar, off the coast of southern Spain. What is more, the welfare of this small population of apes is guarded by the British Army, aided by a subsistence allowance from civil funds.

Barbary apes are not real apes, rather they are monkeys belonging to the macaque family. However, they are called apes because the tail is so short that it is barely visible. In general,

BARBARY APE

CLASS	**Mammalia**
ORDER	**Primates**
FAMILY	**Cercopithecidae**
GENUS AND SPECIES	***Macaca sylvanus***

ALTERNATIVE NAME
Barbary macaque

WEIGHT
25–33 lb. (11–15 kg)

LENGTH
Head and body: 2–2⅓ ft. (60–70 cm);
tail: 1–2 in. (2.5–5 cm)

DISTINCTIVE FEATURES
Large monkey with no obvious tail; coat is
thick, coarse and yellowish brown in color;
naked, pink face

DIET
Mainly fruits; also seeds, flowers, pine
cones, leaves and buds; some insects,
scorpions and other small animals

BREEDING
Age at first breeding: 3–4 years; breeding
season: year-round, but mainly in winter;
number of young: 1; gestation period: 180
days; breeding interval: 1 year

LIFE SPAN
Up to 25 years in captivity

HABITAT
Gibraltar: rocky cliffs and scrub; Morocco
and Algeria: high forests in mountainous
areas at up to 6,500 ft. (2,000 m)

DISTRIBUTION
Gibraltar; High Atlas Mountains of
Morocco and Algeria

STATUS
Vulnerable; population: 14,000 to 23,000,
75 percent in Morocco

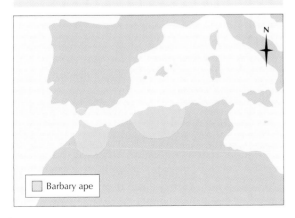

Barbary ape

curiosity leads them to investigate, and sometimes steal, any object that interests them. As they
move about feeding, the monkeys post sentries to
keep a look out and guard the band.

Hairless babies

There is no particular breeding season and the
young barbary apes may be born at any time of
year. However, most births take place during the
summer after a gestation period of about 180
days, a little longer than for other macaques. The
babies are almost hairless at birth and are nursed
by the mother for about 1 year. The father is
known to help care for the baby on occasions.

Props of an empire

The continued existence of the barbary ape on
Gibraltar is due to a superstition that sprang up
when the French and Spaniards attacked the Rock
from 1779–1783. It was said that if the apes left the
Rock, then the British would lose control of the
colony. As a result, the barbary apes are accorded
the same care and protection as the ravens of the
Tower of London, which are the subject of a
similar legend. During World War II the number
of apes dwindled to such a degree that in 1942
Winston Churchill, Britain's Prime Minister at the
time, sent a cable ordering the population to be
maintained at all costs. Additional monkeys were
ferried over from the mainland of North Africa
and the new blood they introduced contributed to
today's larger, healthier colony.

*Barbary apes are the
only wild monkey
remaining in Europe
today, and are also
the only macaques
found in Africa.*

BARNACLE

ARNACLES ARE COMMON sea creatures that encrust rocks, shells, the piles of piers and the bottoms of ships. They are often mistaken for mollusks, as they bear a resemblance to limpets and mussels. However, they are in fact crustaceans, related to amphipods, shrimps and lobsters.

The barnacle's head is firmly cemented to rock or timber by secretions from the first pair of antennae, or antennules. The body is enclosed in five calcareous (limy) plates that give the barnacle its superficially mollusklike character. Within the cavity formed by these plates and the body lie six pairs of forked limbs. These limbs, or cirri, are fringed with stiff hairs, or setae, and are equivalent to the limbs used for walking or swimming by other crustaceans. The cirri beat in unison, flicking in and out through a gap between the plates. Food is collected on the setae or is drawn in by the current set up by the beating of the cirri. This beating also circulates water with dissolved oxygen around the gills for respiration.

Millions on every shore

Acorn barnacles are the most numerous animals on the shore, clustering in groups on rocks, stones and shells. Figures of 30,000 per square yard (25,000 per sq m) are often quoted.

There are two major types of barnacles. The acorn barnacles resemble small limpets, but the shell is made up of several plates. The common acorn barnacle, *Semibalanus balanoides*, of many northerly shores, attains a diameter of nearly ½ inch (1.3 cm), while the American barnacle, *Balanus nubilis*, reaches a diameter of nearly 1 foot (30 cm). Acorn barnacles extend well up the shore, as well as living on ships, and may be exposed to the air for a considerable time each tide. A muscle running across the body contracts to pull the plates together, preventing the barnacle from losing water when exposed. This gives the impression of it having a solid shell. When the tide floods back, the plates open and the feathery limbs commence their continuous clutching for food.

The other main type of barnacle, the goose barnacles, such as *Lepas anatifera*, hang from a tough stalk that is formed from the front part of the head. They normally live on floating timbers, buoys and ship's bottoms. They are also found on flotsam such as driftwood and bottles. Some species form a mucus bubble that hardens to act as a float, from which several individuals may hang. Both acorn and goose barnacles also live on animals such as turtles and whales. One acorn barnacle, genus *Coronula*, is commonly found on whales. It reaches a diameter of 3 inches (8 cm).

Barnacles can be a considerable pest, as a layer of barnacles on a ship's bottom severely reduces its speed. From early times it has been common practice to ground ships and careen them, scraping the encrustations off from below the waterline. Nowadays fouling by barnacles is reduced by special paints that deter the larvae from settling.

Filter feeders

Like other sedentary animals, barnacles can do little more than wait for food to come to them. They generally settle where there is likely to be a current of water passing over them that will carry fresh food for the cirri to sweep up. As the cirri sweep back into the shell they fold over so any captured prey is firmly trapped. The food is then swept down toward the mouth, where it is wiped off the cirri by the appendages surrounding the mouth. Edible matter is pushed into the mouth, while anything inedible is pushed away.

Small creatures up to 1 millimeter long can be caught in the net of the cirri. These include the larvae of other crustaceans and of other marine organisms. Small organisms such as single-celled planktonic plants are also caught and eaten, as are bacteria.

A barnacle actively feeding by using its frilled limbs, or cirri, to capture small marine organisms, sweeping them down into its mouth.

BARNACLES

PHYLUM	**Crustacea**
CLASS	**Maxillopoda**
SUBCLASS	**Cirripedia**
ORDER	**Thoracica**

GENUS AND SPECIES **Common acorn barnacle,**
Semibalanus balanoides; **American barnacle,**
Balanus nubilis; **goose barnacle,** *Lepas*
anatifera; **many others**

LENGTH
**Common acorn barnacle: ⅕–1 in. (0.5–2.5 cm).
American barnacle: up to 1 ft. (30 cm).
Goose barnacle: ¾–4 in. (2–10 cm).**

DISTINCTIVE FEATURES
**Acorn barnacles: conical shell consisting of
5 calcareous (limy) plates; feathery
appendages grab particles from water.
Goose barnacles: long stalk with the
(somewhat reduced) calcareous plates at
unattached end.**

DIET
Plankton of various sizes

BREEDING
**Age at first breeding: 3 years; breeding
season: usually year-round; number of
eggs: 400 to 8,000; hatching period: about
4 months; larva molts several times before
changing into adult form**

LIFE SPAN
**Acorn barnacle: up to 6 years; other species
sometimes longer than this**

HABITAT
**Rocky, intertidal shores; flotsam and ship
bottoms; subtidal rocks, shells and stones;
some species on whales and turtles**

DISTRIBUTION
Worldwide

STATUS
Abundant

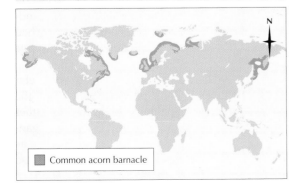

Common acorn barnacle

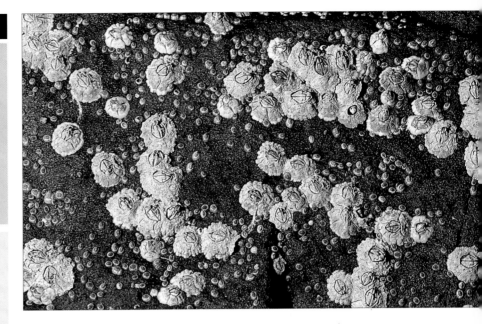

Hermaphroditic reproduction

Barnacles are hermaphrodites, each individual
having both male and female organs. Fertilization
takes place within the cavity formed by the plates
and the body. In the case of the common acorn
barnacle, the eggs are fertilized by sperm cells of a
neighboring individual, which unrolls a 1½-inch
(4-cm) long penis through the opening of its shell
and protrudes it into one of its neighbors.
Development, of the egg takes about 4 months, but
the larvae are not released from the parent until
conditions are favorable, when the adults secrete a
so-called hatching substance. This is secreted when
the adults are well fed, so that larvae will be
released when there is an adequate supply of food.

The newly hatched larva is a minute creature
called a nauplius. It has a round head tapering to
a spiny tail and three pairs of limbs. The
nauplius larva feeds and grows, molting several
times and eventually changing into a cypris
larva, resembling a minute, two-shelled shellfish.
The cypris larva has six pairs of limbs. It does not
feed but drifts with the tide, feeling around for a
suitable place to settle. Then it anchors itself by
its antennules and rapidly changes into the adult
form. The calcareous plates are secreted and the
limbs change from walking legs to waving cirri.
Choosing a place to settle is very important, for
once adult, a barnacle is unable to move for the
rest of its life, which may be several years.

Prized food

Despite the protection of their shells, barnacles
have a number of predators, notably gastropods
such as dog-whelks, but also starfish, crabs and
small fish. One goose barnacle, *Pollicipes pollicipes*,
found in wave-battered crevices on exposed
shores in Spain and Portugal, is so prized as a
food that people will risk life and limb to collect it.

*Common acorn
barnacles at low tide.
A barnacle's body is
enclosed in five plates,
drawn together to
prevent it losing water
when exposed to the air.*

BARNACLE GOOSE

I N AUTUMN AND WINTER, the barnacle goose comes south from its haunts in the Arctic. Smaller than the more common greylag and Canada geese, it is readily recognized by its black-and-white plumage, the forehead and face being a conspicuous white, with black between the eyes and bill. At a distance the barnacle goose can be distinguished from the Canada goose by its lavender gray and black-striped upperparts and grayish white belly, contrasting with the gray-brown of the Canada goose. Its white face distinguishes it from the otherwise somewhat similar plumage of the brent goose. Its legs and the small bill are black.

The barnacle goose is 2–2⅓ feet (60–70 cm) from head to tail, with a wingspan up to 4⅗ feet (1.5 m). It can weigh up to 5 pounds (2.2 kg).

Mystery bird

In the Middle Ages it was thought that the barnacle goose hatched from ship barnacles, and was therefore "fish" rather than "fowl" so could be eaten as food on Fridays. It is only in the past 100 years or so that the barnacle goose has ceased to be a mystery bird. Prior to this no one knew where it bred. Every year, usually in October or

It is only in the last 100 years or so that people have fully understood the migration and breeding patterns of the barnacle goose.

November, flocks would appear around the coasts of Ireland and in the west of Scotland, and also around Holland and Denmark. Occasionally they would be seen elsewhere around the coasts of Europe, and there are a few records of visitors to the eastern seaboard of North America. Dramatically, in mid-April or May, they would disappear again. It was only in 1907 that they were found breeding in Spitzbergen, north of Norway. A year later breeding colonies were also found in northeastern Greenland. In addition, they are now known to breed on Novaya Zemlya, a large island lying north of Russia, and there is also a small population in Iceland.

The nesting sites are usually on cliffs or rocky outcrops, where the geese are safe from attacks by Arctic foxes. When not attending to the nest, the barnacle goose frequents marshes and river valleys. In the winter it inhabits pastures and marshlands, rarely going inland, occasionally moving onto tidal mudflats.

Grass eaters

The general behavior of barnacle geese is the same as that of other geese. They are extremely gregarious, and sometimes even mix with other species of geese in small parties to feed. They are mostly nocturnal and occasionally go to sea, settling on calm water. Barnacle geese feed on a variety of plant material such as mosses, leaves, catkins, seeds and grasses. They also eat some animal matter, for example winkles, worms and various kinds of shrimps. In their winter quarters, however, over 90 percent of their diet is made up of various grasses.

Cliff breeding

Breeding takes place from late May to June. The nests are made of down (the fluffy, insulating layer of feather under the main plumage), together with fragments of moss and lichen. The cup-shaped nests are used year after year and are usually found on small, rocky ledges, sometimes at a considerable height up a cliff face. This is a strange nesting place for so large a bird; most other goose species nest on fairly level ground. The pink-footed goose, however, is another cliff nester. The clutch is usually of